LOW-FREQUENCY NOISE IN ADVANCED MOS DEVICES

ANALOG CIRCUITS AND SIGNAL PROCESSING SERIES
Consulting Editor: Mohammed Ismail. *Ohio State University*

Titles in Series:
CMOS SINGLE CHIP FAST FREQUENCY HOPPING SYNTHESIZERS FOR WIRELESS MULTI-GIGAHERTZ APPLICATIONS
Bourdi, Taoufik, Kale, Izzet
ISBN: 978-1-4020-5927-8

ANALOG CIRCUIT DESIGN TECHNIQUES AT 0.5V
Chatterjee, S., Kinget, P., Tsividis, Y., Pun, K.P.
ISBN-10: 0-387-69953-8

IQ CALIBRATION TECHNIQUES FOR CMOS RADIO TRANCEIVERS
Chen, Sao-Jie, Hsieh, Yong-Hsiang
ISBN-10: 1-4020-5082-8

FULL-CHIP NANOMETER ROUTING TECHNIQUES
Ho, Tsung-Yi, Chang, Yao-Wen, Chen, Sao-Jie
ISBN: 978-1-4020-6194-3

THE GM/ID DESIGN METHODOLOGY FOR CMOS ANALOG LOW POWER INTEGRATED CIRCUITS
Jespers, Paul G.A.
ISBN-10: 0-387-47100-6

PRECISION TEMPERATURE SENSORS IN CMOS TECHNOLOGY
Pertijs, Michiel A.P., Huijsing, Johan H.
ISBN-10: 1-4020-5257-X

CMOS CURRENT-MODE CIRCUITS FOR DATA COMMUNICATIONS
Yuan, Fei
ISBN: 0-387-29758-8

RF POWER AMPLIFIERS FOR MOBILE COMMUNICATIONS
Reynaert, Patrick, Steyaert, Michiel
ISBN: 1-4020-5116-6

ADVANCED DESIGN TECHNIQUES FOR RF POWER AMPLIFIERS
Rudiakova, A.N., Krizhanovski, V.
ISBN 1-4020-4638-3

CMOS CASCADE SIGMA-DELTA MODULATORS FOR SENSORS AND TELECOM
del Río, R., Medeiro, F., Pérez-Verdú, B., de la Rosa, J.M., Rodríguez-Vázquez, A.
ISBN 1-4020-4775-4

SIGMA DELTA A/D CONVERSION FOR SIGNAL CONDITIONING
Philips, K., van Roermund, A.H.M.
Vol. 874, ISBN 1-4020-4679-0

CALIBRATION TECHNIQUES IN NYQUIST A/D CONVERTERS
van der Ploeg, H., Nauta, B.
Vol. 873, ISBN 1-4020-4634-0

ADAPTIVE TECHNIQUES FOR MIXED SIGNAL SYSTEM ON CHIP
Fayed, A., Ismail, M.
Vol. 872, ISBN 0-387-32154-3

WIDE-BANDWIDTH HIGH-DYNAMIC RANGE D/A CONVERTERS
Doris, Konstantinos, van Roermund, Arthur, Leenaerts, Domine
Vol. 871 ISBN: 0-387-30415-0

METHODOLOGY FOR THE DIGITAL CALIBRATION OF ANALOG CIRCUITS AND SYSTEMS: WITH CASE STUDIES
Pastre, Marc, Kayal, Maher
Vol. 870, ISBN: 1-4020-4252-3

HIGH-SPEED PHOTODIODES IN STANDARD CMOS TECHNOLOGY
Radovanovic, Sasa, Annema, Anne-Johan, Nauta, Bram
Vol. 869, ISBN: 0-387-28591-1

LOW-POWER LOW-VOLTAGE SIGMA-DELTA MODULATORS IN NANOMETER CMOS
Yao, Libin, Steyaert, Michiel, Sansen, Willy
Vol. 868, ISBN: 1-4020-4139-X

DESIGN OF VERY HIGH-FREQUENCY MULTIRATE SWITCHED-CAPACITOR CIRCUITS
U, Seng Pan, Martins, Rui Paulo, Epifânio da Franca, José
Vol. 867, ISBN: 0-387-26121-4

DYNAMIC CHARACTERISATION OF ANALOGUE-TO-DIGITAL CONVERTERS
Dallet, Dominique; Machado da Silva, José (Eds.)
Vol. 860, ISBN: 0-387-25902-3

ANALOG DESIGN ESSENTIALS
Sansen, Willy
Vol. 859, ISBN: 0-387-25746-2

LOW-FREQUENCY NOISE IN ADVANCED MOS DEVICES

by

Martin von Haartman and Mikael Östling
KTH, Royal Institute of Technology, School of Information and Communication Technology, Kista, Sweden

A C.I.P. Catalogue record for this book is available from the Library of Congress.

ISBN 978-1-4020-5909-4 (HB)
ISBN 978-1-4020-5910-0 (e-book)

Published by Springer,
P.O. Box 17, 3300 AA Dordrecht, The Netherlands.

www.springer.com

Printed on acid-free paper

All Rights Reserved
© 2007 Springer
No part of this work may be reproduced, stored in a retrieval system, or transmitted
in any form or by any means, electronic, mechanical, photocopying, microfilming, recording
or otherwise, without written permission from the Publisher, with the exception
of any material supplied specifically for the purpose of being entered
and executed on a computer system, for exclusive use by the purchaser of the work.

This book is dedicated to Anne

Contents

Authors	ix
Preface	xi
Acknowledgments	xv
Chapter 1 – Fundamental noise mechanisms	1
Chapter 2 – Noise characterization	27
Chapter 3 – $1/f$ noise in MOSFETs	53
Chapter 4 – $1/f$ noise performance of advanced CMOS devices	103
Chapter 5 – Introduction to noise in RF/analog circuits	175
Appendix I – List of Symbols	189
Appendix II – List of Acronyms	197
Appendix III – Solutions to problems	199
Index	211

Authors

Martin von Haartman and Mikael Östling
KTH, Royal Institute of Technology, School of Information and Communication Technology, Kista, Sweden

PREFACE

The excess noise above the well-known thermal noise and shot noise that shows up at low frequencies, the so-called *low-frequency noise* (other names are $1/f$ noise or flicker noise), has raised questions for a long time and has now become more important than ever. The low-frequency noise generated in the electronic devices is a key problem in analog circuits and systems since it sets a limit on how small signals that can be detected and processed in the circuits. In the early 1990s, the metal-oxide-semiconductor field-effect-transistor (MOSFET) had a channel length of around 0.5 μm and was mainly used in digital electronics. The MOS transistor at that time had a conventional Si channel, SiO_2 gate dielectrics and few advanced features. The tremendous improvements in CMOS performance during the last decade, resulting from continuous advances in the CMOS technology, have stimulated the recent explosion in information and communication technology. Nowadays, MOS transistors are not used only in digital applications but also in a wide range of analog circuits. The low frequency noise in the CMOS devices has therefore emerged as an important concern. The rapid shrinking of the device dimensions (the smallest gate length is around 30 nm in 2006) is not only a challenging technological problem, the low-frequency noise also increases as the dimensions become smaller with fewer and fewer charge carriers in the active region of the device. It has even been predicted that low-frequency noise will be a problem in digital applications in a few years time.

The CMOS technology has also evolved from the standard Si/SiO_2 material system to more advanced material combinations and new types of device structures. This technology shift has had a pronounced impact on the low-frequency noise properties. The introduction of high-k materials or other advanced features accompanied with more complex fabrication processes often lead to (more) defects and imperfections in the current path, which can cause a severe degradation of the low-frequency noise performance. A thorough understanding of the low-frequency noise mechanisms, potential noise sources, various noise models, and the impact of technology are thus important for professionals, researchers and students in the electronics field. In particular those working with CMOS device technology and design, characterization and modeling, and circuit design are expected to find great use of this book. The low-frequency noise cannot be completely eliminated, but with careful design of the devices and clever utilization or development of the technology the low-frequency noise can be substantially reduced. Accurate characterization and modeling of the low-frequency noise is not only immensely important for analog circuit designers but also in order to provide an understanding of the noise phenomenon itself. Furthermore, with deeper insights on how the low-frequency noise affects the output noise of a circuit, ways to optimize the circuit for low noise can be sought out.

This book spans from fundamental noise theory via characterization, MOSFET noise models and CMOS technology to address noise in analog/RF circuits. The purpose is both to give the reader an in-depth knowledge of low-frequency noise, while still presented in an easily comprehensible form, and bring together the different pieces all the way from the fundamental theories and physics level to the circuit level. The focus is on MOS devices and technology but the first two chapters about fundamental noise mechanisms and low-frequency noise characterization provide a general background. Other types of FET devices than the MOSFET, bipolar transistors or devices in other materials than Si/SiGe are beyond the scope of this book and are not treated in detail.

This book is structured as follows. In chapter 1, we will give an introduction to noise, describing the fundamental noise sources and basic circuit analysis. The characterization of low-frequency noise is discussed in detail in chapter 2. We will describe the equipment, measurement setups and diagnostic techniques including many useful practical advices. The various theoretical and compact low-frequency ($1/f$) noise models in MOS transistors are treated extensively in chapter 3, providing an in-depth understanding of the low-frequency noise mechanisms and the potential sources of the noise in MOS transistors. We will give an introduction to the MOS transistor and present its noise equivalent circuit. The number and mobility fluctuation noise models are discussed in detail and the $1/f$ noise dependence on device parameters and operating conditions are explained. We also review the most popular compact noise models; the SPICE and Berkeley short channel IGFET (BSIM3) models. In chapter 4, a comprehensive overview of state-of-the-art CMOS technology is presented together with an exhaustive investigation of the low-frequency noise properties in the various types of advanced CMOS devices. Our presentation includes nanometer scaled devices, strained Si, SiGe, SOI, high-k gate dielectrics, metal gates and finally multiple gates. The book ends with an introduction to noise in analog/RF circuits and describes how the low-frequency noise can affect these circuits. We particularly discuss the voltage controlled oscillator and the upconversion of $1/f$ noise to phase noise as well as the noise properties of mixers and low-noise voltage amplifiers. In order to enhance the understanding of the various aspects of noise fundamentals and the noise implications in advanced CMOS technology, we have composed a number of relevant problems after each chapter. In appendix III a short solution manual is provided.

A reader of this book is assumed to understand fundamental semiconductor physics as well as the principles of CMOS devices at an undergraduate level. Knowledge about noise, CMOS device fabrication or electrical circuits is useful but not necessary. We have mainly followed the conventional notations used in for example Fundamentals of Modern VLSI Devices by Y. Taur & T. K. Ning (Cambridge University Press, Cambridge, 1998). Note that the words low-frequency noise and $1/f$ noise are both frequently used throughout this book, but their meaning is interchangeable for the most part.

Martin von Haartman and Mikael Östling

January 2007

ACKNOWLEDGMENTS

The material of this book mainly stems from the PhD thesis work performed by Dr. Martin von Haartman under the supervision by Prof. Mikael Östling. The idea of the book was born in connection with the PhD defense, for this we would like to thank Prof. Mohammed Ismail. The actual writing of the book was made possible by faculty funds arranged by Prof. Östling during a 6 months assignment as research associate for Martin von Haartman. As always, the realization of this project would never have become possible without the direct and indirect help and contributions from a number of people. The researchers and engineers at the device technology laboratory at KTH (Royal institute of Technology), Kista, are greatly acknowledged for preparing the devices used in this work as well as enlightening research discussions. In this context, we specifically would like to thank Dr. Gunnar Malm for continuous support in form of comments and ideas about this research work as well as Dr. Per-Erik Hellström for the development of the CMOS device technology at KTH fuelling our research work with advanced devices to study. Dr. Malm and Prof. Carl-Mikael Zetterling also deserve many thanks for their help proofreading the manuscript. The authors are very grateful to the continuous project funding through the Swedish foundation for strategic research (SSF), the Swedish Governmental Agency for Innovation Systems (VINNOVA) and the graduate student fellowship award by IEEE EDS. Finally, Martin von Haartman would like to express his deep love and gratitude to his wife Anne for her support and encouragement during the course of this demanding project.

Chapter 1

FUNDAMENTAL NOISE MECHANISMS

1. INTRODUCTION

Currents and voltages in an electronic circuit are perturbed from their given values due to interference of noise. The desired signal becomes difficult to distinguish when the noise power is significant in relation to the signal power, why noise is unwanted in electronic systems. One could categorize noise originating from (i) external sources, for example adjacent circuits, AC power lines, radio transmitters disturbing the circuit of interest due to electrostatic and electromagnetic coupling, and (ii) internal random fluctuations in physical processes governing the electron transport in a medium. This book deals exclusively with the latter type of noise, true noise, and hereafter all noise is understood to be of this type. Due to its random nature, the noise cannot be completely eliminated and therefore ultimately limits the accuracy of measurements and sets a lower limit on how small signals that can be detected and processed in an electronic circuit.[1,2] Thus, noise is a fundamental problem in science and engineering, important to understand, characterize and consider in order to be able to minimize its effects and estimate the accuracy of detected signals.

This chapter begins with a background to noise, how it is defined and the mathematics involved. The fundamental noise mechanisms, thermal noise, generation-recombination noise, random-telegraph signal noise and $1/f$ noise, are discussed in section 3. The analysis of circuits including noise sources is presented in section 4.

2. BASIC NOISE THEORY

2.1 Noise definition

True noise in an electronic device is a random, spontaneous perturbation of a deterministic signal inherent to the physics of the device. Disturbances in an electronic system originating from external sources, for example crosstalk between adjacent circuits, vibrations, light, interference from AC power lines, radio transmitters etc are not considered as noise in this work, as mentioned before. These external disturbances can often be eliminated by appropriate shielding, filtering and layout design of the circuits. True noise, on the other hand, cannot be eliminated, but it is possible to reduce it by proper design of the devices and circuits.

Fig. 1-1 illustrates how an electronic signal fluctuates randomly due to noise. The current through a device can be written as

$$I(t) = \overline{I} + i_n(t) \tag{1-1}$$

where $\overline{I}$ is the average bias current and $i_n(t)$ is a randomly fluctuating current. The value of i_n is random at any point in time and cannot be predicted. Instead we describe the noise with averages, for example the average of i_n measured over a long time should always equal zero. The study of noise is built on the mathematical methods from probability theory, which allows us to define appropriate averages for the random variables we are dealing with. A common and powerful method to characterize and describe noise is by converting the problem from the time domain to the frequency domain by Fourier transformation.

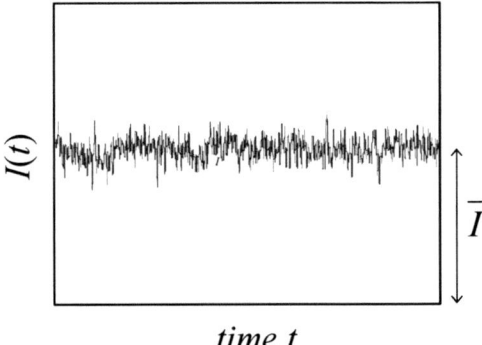

Figure 1-1. A typical noise waveform is illustrated.

1. Fundamental noise mechanisms

2.2 Mathematical treatment

Here we present a brief summary of the most important mathematical methods from probability theory, which serves as a background to the analysis of noise.

Let us consider an electronic circuit and assume for the time being that we have a large number of identical copies of this circuit, an ensemble. At a chosen location in the circuit at a certain point in time there is a probability dP that the wanted signal will be disturbed by noise with an amplitude in the interval [X, X+dX], where X is a random variable. One can define a probability density function $f(X)$ of X and write

$$dP = f(X)dX. \qquad (1\text{-}2)$$

The probability density function should be normalized (scaling of $f(X)$ with a constant) so that the integration over all allowed values of X yields 1. If $f(X)$ is independent of time the random process (also called stochastic process) is said to be stationary, which always is assumed for the noise processes considered in this work.

For random variables, several ensemble averages are defined; mean value, variance, autocorrelation function, power spectral density etc. While one cannot know the exact outcome of random event, these averages give us adequate information about it. The mean value and variance are defined below in Eqs. (1-3) and (1-4), respectively:

$$\overline{X} = \int_{-\infty}^{\infty} X f(X) dX \qquad (1\text{-}3)$$

$$\operatorname{var} X = \overline{(X - \overline{X})^2} = \int_{-\infty}^{\infty} (X - \overline{X})^2 f(X) dX = \overline{X^2} - (\overline{X})^2. \qquad (1\text{-}4)$$

The ensemble averages can be calculated when the probability density function is known. Practically all fluctuating currents and voltages in electrical devices follow the normal (Gaussian) distribution due to the central limit theorem stating that the sum of a large number of independent random variables has a normal distribution. One important exception though is the switching of the signal between two levels, random-telegraph-signal noise, which is a Poisson process. The probability density function for the normal distribution is given as

$$f(X) = \frac{1}{\sigma\sqrt{2\pi}} \exp\left[-\frac{(X-m)^2}{2\sigma^2}\right] \tag{1-5}$$

where $\overline{X} = m$ and var $X = \sigma^2$ if X has a normal distribution. However, the exact probability density function for the noise is seldom known. But, the time averages equal the ensemble averages for certain random processes. That is, the distribution for one ensemble element over time is equal to the distribution over the whole ensemble at a chosen point in time. Such processes are called stationary and *ergodic*. The noise processes discussed here are all considered to be stationary and ergodic, which allows us to use measurements over time for calculation of time averages together with the theory developed for ensemble averages. Currents and voltages are readily measured over time and used to gain information about the noise. The time average of the noise voltage or noise current just equals zero if integrated long enough and provides no interesting information; instead squared quantities are used to describe the noise. One such squared quantity is the power spectral density $S(f)$ which is given from the autocorrelation function $R(s)$ according to the Wiener-Khintchine theorem[3,4]

$$S_x(f) = 4\int_0^\infty R(s)\cos(2\pi f s)ds . \tag{1-6}$$

S_x is a Fourier transformation of $R(s)$, which is given by

$$R(s) = \overline{X(t)X(t+s)} = \lim_{T\to\infty} \frac{1}{T}\int_0^T X(t)X(t+s)dt \tag{1-7}$$

or $R(s) = \int_0^\infty S_x(f)\cos(2\pi f s)df .$ (1-8)

Obviously, if $s = 0$ one obtains the variance or noise "power"

$$\overline{X^2(t)} = \int_0^\infty S_x(f)df = \lim_{T\to\infty}\frac{1}{T}\int_0^T X^2(t)dt . \tag{1-9}$$

The power spectral density (PSD) is measured with a spectrum analyzer, a topic which is discussed in chapter 2. Noise with a constant $S(f)$ for all

1. Fundamental noise mechanisms

frequencies is said to be *white*. It is usually observed that the noise PSD is dependent on frequency at low frequencies, and becomes white thereafter. The corner frequency between frequency dependent noise and white noise is typically from a few Hz up to the MHz range and depends on the type and size of the device, bias conditions etc. A schematic diagram of the PSDs for the excess noise at low frequencies, *low-frequency noise*, and the white noise is shown in Fig. 1-2. The low-frequency noise may consist of superimposed $1/f$ noise (or $1/f$-like noise) and generation-recombination (g-r) noise. The fundamental sources of noise (including the two mentioned above) are discussed in section 3.

Both white noise and low-frequency noise are important to consider in electronic circuits, their relative importance depends on the type of circuit and its application. The physical mechanisms behind the white noise sources are well known and the white noise level can usually be accurately predicted in electronic circuits. However, the origin of the low-frequency noise still raises many questions. For this reason, we have chosen to mainly deal with low-frequency noise in this work.

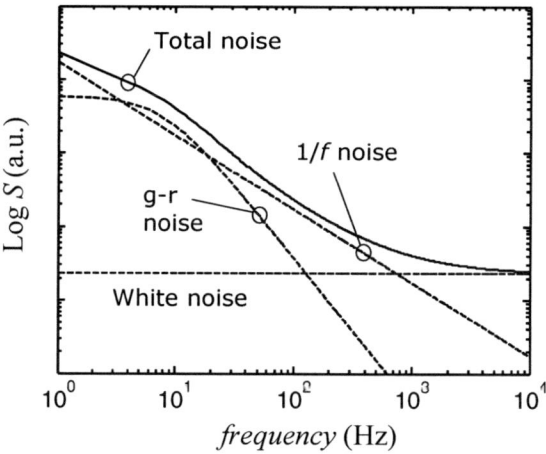

Figure 1-2. The PSD (*S*) for low-frequency noise and white noise plotted vs. frequency. The excess noise above the white noise floor is called low-frequency noise and may consist of $1/f$ noise or generation-recombination (g-r) noise.

2.3 Noise quantities

Here we define some important noise quantities that we will use or are commonly used in the literature. The *power spectral density* that we defined in the previous section gives information about how the noise power is distributed in frequency. The PSD of the noise current and noise voltage has units of A^2/Hz and V^2/Hz, respectively. We use the term *noise power* for the mean square of noise voltages or noise currents and can be thought of as the average power delivered to a 1-Ω resistor within the bandwidth Δf of the system from a fluctuating current or voltage. The *RMS (root mean square)* noise voltage is the square root of the noise voltage power

$$v_{n,rms} = \sqrt{\overline{v_n^2}} = \sqrt{\int_{f_1}^{f_2} S_V df} \approx \sqrt{\frac{1}{T}\int_0^T v_n^2(t)dt} \qquad (1\text{-}10)$$

where v_n is the noise voltage, and S_V is the PSD of the noise voltage and $\Delta f = f_2 - f_1$. The last expression suggests how $v_{n,rms}$ can be measured. Note that the averaging time T should be long enough (some multiples of $1/f_1$).

Quantities such as *noise temperature* and *noise resistance* are sometimes used to indicate the noise level and are defined below

$$T_n = \frac{\overline{v_n^2}}{4kR\Delta f} \qquad (1\text{-}11)$$

$$R_n = \frac{\overline{v_n^2}}{4kT_0\Delta f}. \qquad (1\text{-}12)$$

The temperature $T_0 = 290$ K is the standard noise temperature, $k = 1.38 \times 10^{-23}$ J/K is Boltzmann's constant and R is the resistance. If only thermal noise is present (see next section), then T_n and R_n equal the actual temperature and (ohmic) resistance, respectively. This is true for a metallic resistor for example. But in case other noise sources contribute as well, T_n and R_n are higher than those values.

3. FUNDAMENTAL NOISE SOURCES

There are some fundamental physical processes that can generate the random fluctuations in the current (or voltage) in a device. The current in a conductor is the transported charge through the conductor per unit time. The average current in a slab of length L can be written as

$$\overline{I} = q\overline{Nv_d}/L \qquad (1\text{-}13)$$

1. Fundamental noise mechanisms

where $q = 1.602\times10^{-19}$ C is the electron charge, N is the number of free carriers in the slab and v_d is the drift velocity of the carriers. A bar over a variable always means that the average is taken. Both N and v_d can fluctuate and therefore

$$I(t) = \sum_{i=1}^{N(t)} q \frac{v_i(t)}{L} \tag{1-14}$$

where v_i is the drift velocity for an individual carrier and

$$N(t) = \overline{N} + \Delta N(t) \tag{1-15}$$

$$v_i(t) = \overline{v_i} + \Delta v_i(t). \tag{1-16}$$

For a homogeneous sample subjected to a uniform electric field the average drift velocity is the same for each carrier. The fluctuating current can then be written as

$$\Delta I(t) = \frac{q}{L}\overline{v_d}\Delta N(t) + \frac{q}{L}\sum_{i=1}^{\overline{N}}\Delta v_i(t). \tag{1-17}$$

The first term is due to fluctuating number of carriers and the second term to fluctuating carrier velocity. These are essentially the two sources of noise current fluctuations stemming from physical processes inside a material, but both the carrier number and velocity fluctuations can be generated by different mechanisms. Instead of carrier velocity fluctuations, one can speak of mobility fluctuations. The drift velocity is proportional to the applied electric field E

$$v_d = \mu E \text{ or more general } v_i = \mu_i E. \tag{1-18}$$

The proportionality constant $\mu(\mu_i)$ is the carrier mobility (individual carrier mobility). In the coming subsections, the fundamental sources of noise are discussed and described in terms of the PSD of the noise current.

3.1 Thermal noise

Thermal noise (Nyquist, Johnson noise) stems from the random thermal motion of electrons in a material. Each time an electron is scattered, the velocity of the electron is randomized. Instantly, there could be more electrons moving in a certain direction than electrons moving in the other directions and a small net current is flowing. This current fluctuates in strength and direction, but the average over (long) time is always zero. If a piece of material with resistance R and (non-zero) temperature T is considered, the PSD of the thermal noise current is found to be

$$S_I = \frac{4kT}{R} \text{ (or } S_V = 4kTR\text{)}. \qquad (1\text{-}19)$$

The thermal noise was first discovered experimentally by J. B. Johnson and theoretically explained by H. Nyquist in 1928.[5,6] For this reason, thermal noise is also called Johnson or Nyquist noise. The thermal noise exists in every resistor and resistive part of a device (no bias needs to be applied) and sets a lower limit on the noise in an electric circuit. The thermal noise can, however, not be white up to infinitely high frequencies. Otherwise the noise power could theoretically extend to infinity, which of course is unphysical. It has been shown theoretically that kT in Eq. (1-19) should be replaced by a frequency dependent quantum correction factor[1]

$$S_I = 4\frac{hf}{e^{hf/kT}-1}\frac{1}{R} \qquad (1\text{-}20)$$

where $h = 6.626 \times 10^{-34}$ Js is Planck's constant. At "low" frequencies, $hf \ll kT$, the quantum factor reduces to kT. In practice, this factor is not very important since the bandwidth of the system in reality is smaller than hf and effectively limits the noise power.

The maximum *available noise power* P_n delivered from a conductor to the remaining circuit occurs when the input resistance of the circuit equals the noise generating resistance (matched impedance). The delivered noise power is

$$P_n = \overline{v_n^2}/4R = kT\Delta f \text{ [W]}. \qquad (1\text{-}21)$$

This is a true noise power and is independent of the actual resistance. Of course, no net power transfer takes place since the circuit transfers the same thermal noise power P_n to the conductor. The decibel (dB) scale is often used in communications. Since the dB scale is relative, a reference level must be selected. In the dBm scale, which uses 1 mW as the reference power, the available noise power equals

1. Fundamental noise mechanisms

$$P_{n,dBm} = 10\log_{10}\left(\frac{kT\Delta f}{10^{-3}}\right) = -174 + 10\log_{10}(\Delta f) \text{ [dBm]}. \qquad (1\text{-}22)$$

The thermal noise from resistive elements is unavoidable, but a circuit can be designed in order to minimize it. First of all, reactive elements do not generate thermal noise so input matching techniques using reactive elements can be used to lower the noise in amplifiers. Unused portions of the bandwidth cause unnecessary noise, therefore the system bandwidth should be kept as narrow as possible only to pass the desired signal.

3.2 Shot noise

The current flowing across a potential barrier, like the pn-junction, is not continuous due to the discrete nature of the electronic charge (electrons). The current across a barrier is given by the number of carriers, each carrying the charge q, flowing through the barrier during a period of time. A shot noise current is generated when the electrons cross the barrier independently and at random. The current fluctuates with a PSD

$$S_I = 2qI \qquad (1\text{-}23)$$

where I is the DC current across the barrier. The shot noise is a Poisson process and was first discovered in vacuum tubes by W. Schottky,[7] dating back to 1918. A current is necessary in order to generate shot noise, but the currents crossing the barrier in the backward and forward directions should be considered separately. The (ideal) diode current is given as

$$I = I_0\left(e^{qV_d/kT} - 1\right) \qquad (1\text{-}24)$$

where I_0 is the diode saturation current and V_d is the applied voltage across the pn-junction. The current across a pn-junction under zero bias is zero since the forward (first term in Eq. 1-24) and backward currents cancel each other. However, the total shot noise is the sum of the shot noise in the forward and backward currents

$$S_I = 2qI_0 e^{qV_d/kT} + 2qI_0 = 4qI_0. \qquad (1\text{-}25)$$

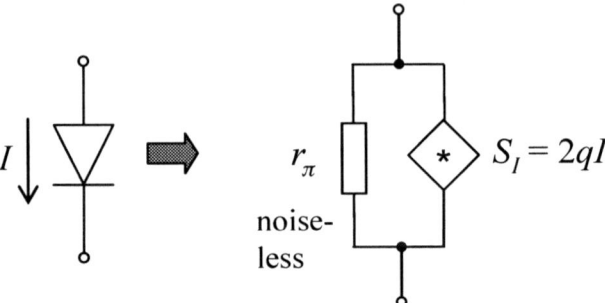

Figure 1-3. Modeling of the shot noise in a diode with a dynamic resistance r_π (=kT/qI) in parallel with a noise current generator delivering $S_I = 2qI$. See further section 4 about noise circuit analysis.

The fact that the sign of the currents are opposite does not matter, the mean square values of the noise are always added positively. A pn-junction is associated with a noise-less dynamic resistance (see Fig. 1-3)

$$r_\pi = \left(\frac{dI}{dV_d}\right)^{-1}. \quad (1\text{-}26)$$

By performing the simple calculation it turns out that the shot noise PSD S_I of the diode equals $4kT/r_\pi$ equivalent to the expected thermal noise of the dynamic resistance if it would act as an ohmic resistance. This shows that the thermal and shot noise phenomena are closely related. When the diode is biased in the forward direction, the forward current is much larger than the backward current ($I \gg I_0$) and $r_\pi = kT/qI$. Thus, $S_I = 2qI = 2kT/r_\pi$ which is half the fictitious thermal noise for the dynamic resistance. The reason behind the factor 1/2 is basically that the current is essentially flowing in one direction across the pn-junction when it is forward biased.

Finally, note that the measured shot noise PSD actually can be lower than $2qI$ if the current pulses across the barrier are correlated. The Fano factor is defined as

$$FF = S_{I,meas}/2qI. \quad (1\text{-}27)$$

Usually, $FF = 1$ but can be lower than 1 for quantum mechanical conditions (such as in mesoscopic devices at temperatures close to 0 K).

3.3 Generation-recombination noise

Generation-recombination (g-r) noise in semiconductors originates from traps that randomly capture and emit carriers, thereby causing fluctuations in the number of carriers available for current transport. If carriers are trapped at some critical spots, the trapped charge can also induce fluctuations in the carrier mobility, diffusion coefficient, electric field, barrier height, space charge region width etc. Electronic states within the forbidden bandgap are referred to as traps, and exist due to the presence of various defects or impurities in the semiconductor and at its surfaces. Transitions of the following forms occur in a semiconductor

(i) free electron + free hole recombine
(ii) free electron + free hole are generated
(iii) free electron + empty trap ⇆ electron bound to trap
(iv) free hole + empty trap ⇆ hole bound to trap.

Note that a trap may be neutral or charged in its empty state. The PSD of the fluctuations in the number of carriers is found to be[1]

$$S_N(f) = 4\overline{\Delta N^2} \frac{\tau}{1+(2\pi f)^2 \tau^2}. \qquad (1\text{-}28)$$

Here, τ is the time constant of the transitions. The shape of the spectrum given by Eq. (1-28) is called a Lorentzian and is illustrated in Fig. 1-5 in the next section. G-r noise is only significant when the Fermi energy level is close, within a few kT, to the trap energy level. Then the capture time τ_c and the emission time τ_e are almost equal. If the Fermi-level is far above or below the trap level, the trap will be filled or empty most of the time and few transitions occur that produce noise. The variance can be expressed as[8]

$$\frac{1}{\overline{\Delta N^2}} = \frac{1}{N} + \frac{1}{N_{T,full}} + \frac{1}{N_{T,empty}} \qquad (1\text{-}29)$$

where $N_{T,full}$ and $N_{T,empty}$ are the average number of full and empty traps, respectively. At the Fermi-level and assuming $N \gg N_T$ ($N_T = N_{T,full} + N_{T,empty}$)

$$\overline{\Delta N^2} = N_T / 4. \qquad (1\text{-}30)$$

Using Eqs. (1-13), (1-17), (1-28) and (1-30) gives

$$S_I = \frac{S_N}{N^2}I^2 = I^2 \frac{N_T}{N^2} \frac{\tau}{1+(2\pi f)^2 \tau^2}. \quad (1\text{-}31)$$

As seen from Eq. (1-31), the PSD is proportional to the number of traps and inversely proportional to the number of carriers squared. In general, the time constant and the relative strength of the traps differ (depends on the trap energy level and spatial position). For a certain distribution of time constants, the PSD becomes proportional to $1/f$ which is discussed in section 3.5.

3.4 Random-Telegraph-Signal (RTS) noise

A special case of g-r noise is the RTS noise (burst, popcorn noise), which is displayed as discrete switching events in the time domain, see Fig. 1-4. If only a few traps are involved, the current can switch between two or more states resembling a RTS waveform due to random trapping and detrapping of carriers. For two-level pulses with equal height ΔI and Poisson distributed time durations in the lower state τ_l and in the higher state τ_h, the PSD of the current fluctuations is derived as[9]

$$S_I(f) = \frac{4(\Delta I)^2}{(\tau_l + \tau_h)\left[(1/\tau_l + 1/\tau_h)^2 + (2\pi f)^2\right]}. \quad (1\text{-}32)$$

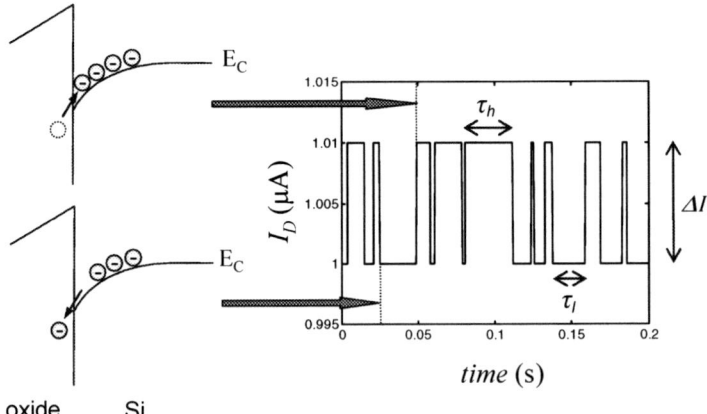

Figure 1-4. Schematic description of RTS noise, exemplified for a MOSFET. The drain current switches between two discrete levels when a channel electron moves in and out of a trap in the gate oxide.

1. Fundamental noise mechanisms 13

The PSD for the RTS noise and the g-r noise are both of the Lorentzian type. Fig. 1-5 shows the Lorentzian PSD for the RTS noise waveform in Fig. 1-4. G-r noise can be viewed as a sum of RTS noise processes from one or more traps with identical time constants, and is only displayed as RTS noise in the time domain if the number of traps involved is small. RTS noise is an interesting phenomenon since the random switching process from just one trap can be studied in the time domain. It is established that RTS noise is caused by a single carrier controlling the flow of a large number of carriers rather than a large number of carriers being involved in the trapping/detrapping process, thus a single electron can be studied. RTS noise and g-r noise are normally very sensitive to the temperature.[10-11] In bipolar and MOS transistors, the bias conditions are also important since the Fermi-level and the carrier density have a strong impact on the noise characteristics. Interesting information about the trap energy level, capture and emission kinetics and spatial location of the trap can be acquired from RTS noise characterizations using temperature or bias dependencies.[11-14]

RTS noise can be observed in MOS devices with a small gate area (usually below 1 μm^2) and/or with low background noise. If a large area device shows RTS noise, the RTS noise is most probably associated with a parasitic current for example at the periphery of the gate. RTS noise is especially sensitive to bottlenecks for the current flow; current crowding or a poor contact could cause RTS noise as well as drastically higher low-frequency noise in general.

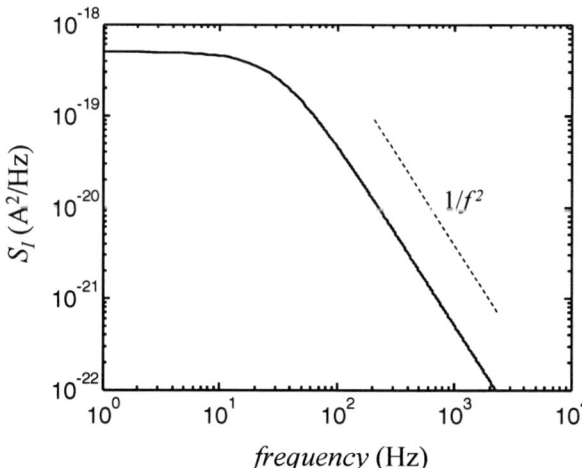

Figure 1-5. A Lorentzian shaped PSD, plotted for the RTS noise waveform in Fig. 1-4.

3.5 1/f noise

1/f noise, also called flicker noise, is the common name for fluctuations with a PSD proportional to $1/f^\gamma$ with γ close to 1, usually in the range 0.7-1.3. The PSD for 1/f noise takes the general form

$$S_I = \frac{K \cdot I^\beta}{f^\gamma} \qquad (1\text{-}33)$$

where K is a constant and β is a current exponent. 1/f fluctuations in the conductance have been observed in the low-frequency part of the spectrum (10^{-5} to 10^7 Hz) in most conducting materials and a wide variety of semiconductor devices.[1,15-17] Analyzing Eqs. (1-17) and (1-18) it is clear that there are essentially two physical mechanisms behind any fluctuations in the current: fluctuations in the mobility or fluctuations in the number of carriers. G-r noise from a large number of traps (number fluctuations) can produce 1/f noise if the time constants of the traps are distributed as[18]

$$g(\tau) = \frac{1}{\ln(\tau_2/\tau_1)\tau} \quad \text{for } \tau_1 < \tau < \tau_2, \quad g(\tau) = 0 \text{ otherwise.} \qquad (1\text{-}34)$$

The factor $1/\ln(\tau_2/\tau_1)$ is for normalization purposes. The superposition of g-r noise from many traps distributed according to $g(\tau)$ yields

$$S_{tot}(f) = \int_0^\infty g(\tau) S_{g-r}(\tau) d\tau = \frac{1}{\ln(\tau_2/\tau_1)} \int_{\tau_1}^{\tau_2} \frac{1}{\tau} \frac{B\tau}{1+(2\pi f\tau)^2} d\tau$$
$$= \frac{1}{\ln(\tau_2/\tau_1)} \frac{B}{2\pi f} \left[\arctan(2\pi f\tau)\right]_{\tau_1}^{\tau_2}. \qquad (1\text{-}35)$$

Thus,

$$S_{tot} \approx \frac{B}{4\ln(\tau_2/\tau_1)f} \quad \text{for } 1/2\pi\tau_2 \ll f \ll 1/2\pi\tau_1. \qquad (1\text{-}36)$$

An example is given in Fig. 1-6 where g-r noise from four individual traps with different time constants adds up to a $1/f^\gamma$ spectrum with γ close to 1. Some remarks are necessary about the addition of g-r noise spectra. First, it is assumed that the g-r noise from the traps can simply be added. This is

1. Fundamental noise mechanisms

true if the traps are isolated and do not interact.[8] Moreover, if the number of carriers is smaller than the effective number of traps, mixing occurs producing g-r noise with a time constant given by the reciprocal sum of all time constants.[19] Secondly, the traps are assumed to couple in the same way to the output current (same B for all traps). Number fluctuation noise is discussed in more detail for the particular case of a MOS transistor in chapter 3.

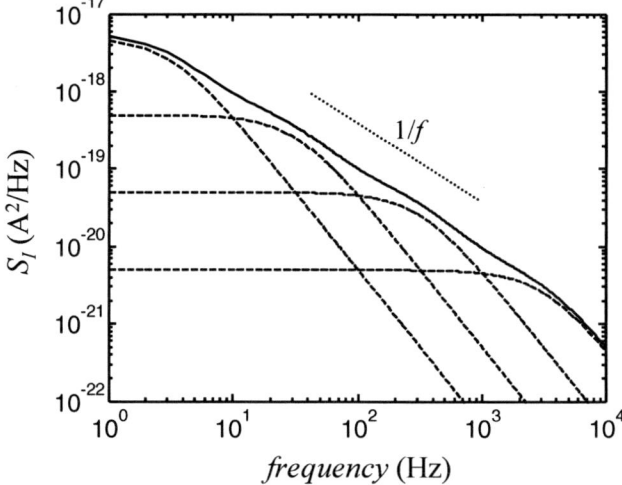

Figure 1-6. Superposition of 4 Lorentzians giving a total spectrum that approximately exhibits a $1/f$ dependence over several decades of frequency.

The second mechanism that can give $1/f$ noise is mobility fluctuations. It was first described by Hooge with the following empirical formula for the resistance fluctuations[20]

$$\frac{S_R}{R^2} = \frac{\alpha_H}{fN}. \qquad (1\text{-}37)$$

The dimensionless parameter α_H, referred to as the Hooge parameter, was first suggested to be constant and equal to 2×10^{-3}. Later, it was found that α_H depends on the crystal quality; in perfect materials 2-3 orders of magnitude lower values were observed. It was also proposed that only phonon scattering contributes to the mobility fluctuations.[21] The factor $1/N$ results from independent mobility fluctuations by each of the N conducting carriers.[22] The conductivity σ in a volume V is given as

$$\sigma = \frac{q}{V}\sum_{i=1}^{N} \mu_i = qN\overline{\mu_i}/V = qn\overline{\mu_i} = qn\mu. \qquad (1\text{-}38)$$

The conductivity fluctuates due to fluctuations in the individual carrier mobilities μ_i (the number fluctuations are not considered here)

$$\Delta\sigma = \frac{q}{V}\sum_{i=1}^{N}\Delta\mu_i. \qquad (1\text{-}39)$$

The fluctuations in the i:th and j:th carriers are assumed to be independent, hence

$$\overline{\Delta\mu_i \cdot \Delta\mu_j} = 0 \text{ for } i \neq j$$

$$\Rightarrow \overline{(\Delta\sigma)^2} = \frac{q^2}{V^2}\sum_{i=1}^{N}\overline{(\Delta\mu_i)^2} = \frac{q^2}{V^2} N \overline{(\Delta\mu_i)^2}. \qquad (1\text{-}40)$$

For the spectral density

$$\frac{S_\sigma}{\sigma^2} = \frac{S_\mu}{\mu^2} = \frac{1}{N}\frac{S_{\mu_i}}{\mu_i^2} = \frac{S_R}{R^2}. \qquad (1\text{-}41)$$

The PSD of the individual mobility fluctuations is then

$$S_{\mu_i}/\mu_i^2 = \alpha_H / f. \qquad (1\text{-}42)$$

which means that α_H is proportional to the variance of the relative mobility fluctuation for each carrier, independent of the number of carriers. The mobility fluctuation noise is always present even without an applied bias, but a bias current facilitates the detection of the noise. The noise in a MOSFET also depends on the bias condition as discussed in chapter 3.

There has been a long debate in the noise research community about the origin of the 1/f noise and two schools of thought (number fluctuation or mobility fluctuations) have emerged. Unfortunately, much of the efforts in the past have been focused on proving either of the two sources as the principal one. In fact, it is likely that both of them contribute and which source that dominates the 1/f noise in a particular situation or is most

1. Fundamental noise mechanisms

important depends on the material, type of device, operating conditions, sample variations etc.

The Hooge model has been successful in explaining the $1/f$ noise in metals and bulk semiconductors.[16] In MOS-transistors, on the other hand, the current is flowing in a narrow path confined close to the surface under the gate oxide. In such case, most evidence point to traps in the gate oxide as the dominant $1/f$ noise source.[14,23,24] Nevertheless, the mobility fluctuation noise model tends to be better to explain the $1/f$ noise in p-channel MOSFETs.[25-27]

The Hooge noise model in Eq. (1-37) is empirical and does not suggest a physical explanation behind the mobility fluctuations. Despite the success of the model, the lack of a theoretical model based on physical principles is a weakness and an annoying circumstance. Several good attempts have been made to develop a theoretical mobility fluctuation noise model, but so far none of them is widely accepted. The most important proposed models are reviewed below.

The disputed quantum noise theory of Handel explains the $1/f$ noise as fluctuations in the electron scattering due to infrared photon emission.[28,29] An electron is decelerated when it is scattered, leading to electromagnetic field radiation, i.e. emission of photons. The photon energy, hf, depends on frequency, resulting in a probability of photon emission proportional to $1/f$ giving the $1/f$ fluctuations in the scattering cross section. The theory has, however, received criticism from both practical and theoretical viewpoints.[30-32] Moreover, the predicted overall Hooge parameter value for Si is in the order of 10^{-8},[33] but reported values for Si MOSFETs range between 10^{-6} and 10^{-3}. Therefore, quantum $1/f$ noise may set a lower limit on the $1/f$ noise, but other sources are likely dominating the $1/f$ noise in the vast majority of devices. Furthermore, the quantum $1/f$ noise theory is difficult to reconcile with the impact of technology on the $1/f$ noise.

Another mobility fluctuation noise theory, proposed recently by Musha and Tacano,[34] suggests that energy partition among weakly coupled harmonic oscillators in an equilibrium system is subjected to $1/f$ fluctuations. The authors derive the relationship $\alpha_H = d/\lambda_e$ where d is the lattice constant and λ_e is the mean free path of the electrons in the case of phonon scattering. It is also worth to mention the theory by Jindal and van der Ziel.[35] They propose that the phonon population exhibits g-r noise which is transferred to mobility fluctuation noise through a fluctuating phonon scattering. The idea is very interesting since it is possible that electrical g-r noise stems from g-r noise in the phonon population. The mobility and number fluctuations might even stem from the same physical mechanism. Mihaila speculates that an inelastic tunneling process involving excitation of phonons is the origin of both the number and mobility fluctuation noise.[36] Just recently, Melkonyan *et al.* proposed a theory explaining the mobility fluctuations due to energy fluctuations resulting from random acoustic phonon-phonon scattering.[37] However, the reader should at this point be cautious regarding the usefulness of these new theories until they have received general acceptance.

4. NOISE CIRCUIT ANALYSIS

This section introduces the analysis of noise sources in electrical circuits. We will constrict ourselves to simple resistive networks; the more advanced case of a MOSFET is discussed in chapter 3. The representation of noise sources in circuits, the addition of noise sources and the analysis of a circuit containing noise sources are presented in the following subsections.

4.1 Representation of noise sources

Electrical noise is small current or voltage fluctuations around a DC value. The small-signal equivalent circuit is therefore appropriate to use for circuit modeling. A noisy resistance is represented with a noiseless resistance in parallel with a noise current generator (Norton equivalent) or in series with a noise voltage generator (Thévenin equivalent), see Fig. 1-7. A resistance always generates thermal noise but may also exhibit superimposed $1/f$ noise. Similarly, other elements like pn-junctions and the channel of a MOS-transistor can be represented by a noiseless element in parallel or in series with a noise generator. We use a circle for a voltage noise source and the diamond symbol for a current noise source, both symbols enclosing a star inside it. It is important to observe that the polarity of the voltage and the direction of the current are random when dealing with noise, which is indicated by the star.

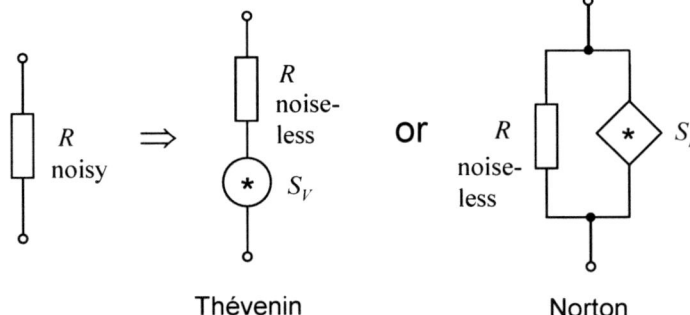

Figure 1-7. Representation of a noisy resistor with Thévenin or Norton equivalent circuits containing a noiseless resistor and a noise generator with PSD S.

4.2 Addition of noise voltages

Consider the simple circuit in Fig. 1-8 consisting of two noise voltage sources in series. The addition of noise currents and voltages is performed using Kirchoff's laws as for normal circuit analysis. For calculation of the mean square total noise, some care must be taken. The average total noise voltage can be calculated as

$$\overline{v_{n,tot}} = \iint (v_{n,1} + v_{n,2}) f(v_{n,1}, v_{n,2}) dv_{n,1} dv_{n,2} = \overline{v_{n,1}} + \overline{v_{n,2}} = 0. \quad (1\text{-}43)$$

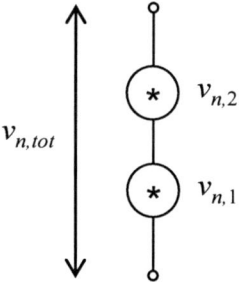

Figure 1-8. Two noise voltage sources in series.

The average noise voltage for both sources equals zero, then the average total noise voltage is also zero (as expected). The total noise power

$$\overline{v_{n,tot}^2} = \iint (v_{n,1} + v_{n,2})^2 f(v_{n,1}, v_{n,2}) dv_{n,1} dv_{n,2}$$
$$= \overline{v_{n,1}^2} + \overline{v_{n,2}^2} + 2\overline{v_{n,1} v_{n,2}} \quad (1\text{-}44)$$

is the sum of the separate noise powers and a cross-product term describing the *correlation* between the two noise sources. The correlation between two random variables can be expressed with the correlation coefficient ρ

$$\rho_{1,2} = \frac{\overline{v_{n,1} v_{n,2}}}{\sqrt{\overline{v_{n,1}^2}} \sqrt{\overline{v_{n,2}^2}}} \quad (1\text{-}45)$$

which can assume values between 1 and -1. It is often the case that two noise sources are *uncorrelated*, then $\rho = 0$. Thus for uncorrelated noise sources, the total noise power or PSD is found by simply adding the noise powers or PSDs for the separate noise sources. Remember that *independent* random variables are always uncorrelated (the opposite is however not necessarily true).

4.3   Simple resistive networks

In order to solve for currents and voltages in circuits containing noise sources, the familiar methods from electrical circuit theory can be employed. We solve for the circuit with noise sources and without noise sources separately; any independent DC or AC source can be zeroed in the noise equivalent circuit. When zeroed, a current source is replaced by an open circuit and a voltage source by a short circuit. For simple circuits containing no dependent sources, the *superposition* principle is often found to be useful. The superposition principle states that the total response in a linear circuit is the sum of the responses for each source acting alone with the other sources zeroed. For circuits containing non-linear dependent sources the superposition principle is not valid, but is advisable to use other methods for circuits with dependent sources even if they are linear. When using the superposition principle in a circuit with uncorrelated noise sources, the noise power contributions from each source can simply be added. Another method which might seem more straightforward but also somewhat more tedious is to

1. assume a direction of the noise current sources and a polarity of the voltage noise sources
2. write and solve circuit equations using Kirchoff's laws
3. group together terms containing the same noise voltage or noise current
4. calculate the noise power (or PSD) according to Eq. (1-44)

We are going to describe how noise circuit analysis is performed by looking at an example. Consider the circuit in Fig. 1-9(a). Both resistors generate thermal noise; thus we replace the noisy resistors with their noise (Thévenin) equivalents. The noise equivalent circuit is presented in Fig. 1-9(b). We first use the superposition principle to solve for the noise current i_n and the noise voltage $v_{n,o}$,

$$i_n = i_{n,1} + i_{n,2} = \frac{v_{n,1}}{R_1 + R_2} - \frac{v_{n,2}}{R_1 + R_2} \qquad (1\text{-}46)$$

1. Fundamental noise mechanisms

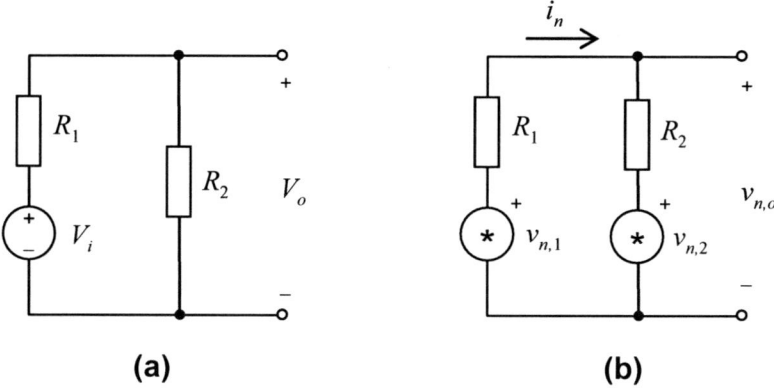

Figure 1-9. Noise equivalent circuit (b) of the circuit in (a).

and

$$v_{n,o} = \frac{v_{n,1} R_2}{R_1 + R_2} + \frac{v_{n,2} R_1}{R_1 + R_2}. \tag{1-47}$$

Since the thermal noise in R_1 and R_2 is uncorrelated, the noise powers are given as

$$\overline{i_n^2} = \frac{\overline{v_{n,1}^2}}{(R_1 + R_2)^2} + \frac{\overline{v_{n,2}^2}}{(R_1 + R_2)^2} \tag{1-48}$$

$$\overline{v_{n,o}^2} = \frac{\overline{v_{n,1}^2} R_2^2}{(R_1 + R_2)^2} + \frac{\overline{v_{n,2}^2} R_1^2}{(R_1 + R_2)^2}. \tag{1-49}$$

When using the superposition principle, each term represent the noise response for one noise source only. Then the latter two equations can be written directly without the intermediate step in Eqs. (1-46) and (1-47).

If we instead write $v_{n,o}$ as

$$v_{n,o} = i_n R_2 + v_{n,2}. \tag{1-50}$$

The operation

$$\overline{v_{n,o}^2} = \overline{i_n^2} R_2^2 + \overline{v_{n,2}^2} \qquad (1\text{-}51)$$

is *wrong!* This is because i_n is correlated with $v_{n,2}$ since the noise current is generated partly from $v_{n,2}$. We must group together all terms containing $v_{n,1}$ and $v_{n,2}$ before we can square the terms

$$v_{n,o} = i_n R_2 + v_{n,2} = \frac{v_{n,1} R_2 - v_{n,2} R_2}{R_1 + R_2} + v_{n,2} = \frac{v_{n,1} R_2 + v_{n,2} R_1}{R_1 + R_2}. \qquad (1\text{-}52)$$

Now the correct noise power can be calculated.

SUMMARY

- Noise is a random phenomenon that causes fluctuations of the currents and voltages in electrical circuits.
- Due to the randomness, noise is described by averages such as the power spectral density, mean square (noise "power") or root mean square of the fluctuating quantity.
- There are five fundamental noise sources; 1) thermal noise, 2) shot noise, 3) generation-recombination (g-r) noise, 4) RTS noise (special case of g-r noise) and 5) $1/f$ noise.
- Thermal noise and shot noise are white noise sources, i.e. their PSDs are frequency independent.
- Low-frequency noise is the excess noise above the white noise level at low-frequencies. The PSD of low-frequency noise increases as the frequency decreases, if the PSD is roughly proportional to $1/f$ one speaks of $1/f$ noise or flicker noise.
- Low-frequency noise is especially troublesome in devices or sensors operating at low frequencies, but is also upconverted to RF frequencies in voltage controlled oscillators (see chapter 5).
- For *uncorrelated* noise sources, the total PSD or noise power is the sum of the PSD or noise power for each source. For two *correlated* sources, a cross-product term describing the *correlation* between the two noise sources must be added.
- A noisy resistor is represented by a Thévenin or Norton equivalent circuit containing a noiseless resistor and a noise generator.

REFERENCES

1. A. van der Ziel, *Noise in solid state devices and circuits* (John Wiley & Sons, New York, 1986).
2. C. D. Motchenbacher and J. A. Connelly, *Low-noise electronic system design* (John Wiley & Sons, New York, 1993).
3. N. Wiener, Generalized harmonic analysis, *Acta Math.* **55**, 117 (1930).
4. A. Khintchine, Korrelationstheorie der stationären stochastischen prozesse, *Math. Ann.* **109**, 604 (1934).
5. J. B. Johnson, Thermal agitation of electricity in conductors, *Phys. Rev.* **32**, 97-109 (1928).
6. H. Nyquist, Thermal agitation of electric charge in conductors, *Phys. Rev.* **32**, 110-113 (1928).
7. W. Schottky, Über spontane Stromschwankungen in verschiedenen Elektrizitätsleitern, *Annalen der Physik* **57**, 541-567 (1918).
8. F. N. Hooge, $1/f$ noise sources, *IEEE Trans. Electron Devices* **41**, 1926-1935 (1994).
9. S. Machlup, Noise in semiconductors: spectrum of a two-parameter random signal, *J. Appl. Phys.* **25**, 341-343 (1954).
10. G. Bosman and R. J. J. Zijlstra, Generation-recombination noise in p-type silicon, *Solid-State Electron.* **25**, 273-280 (1982).
11. M. J. Kirton and M. J. Uren, Noise in solid-state microstructures: a new perspective on individual defects, interface states and low-frequency ($1/f$) noise, *Advances in Physics* **38**, 367-468 (1989).
12. N. V. Amarasinghe, Z. Çelik-Butler, and A. Keshavarz, Extraction of oxide trap properties using temperature dependence of random telegraph signals in submicron metal-oxide-semiconductor field-effect transistors, *J. Appl. Phys.* **89**, 5526-5532 (2001).
13. M. von Haartman, M. Sandén, M. Östling, and G. Bosman, Random telegraph signal noise in SiGe heterojunction bipolar transistors, *Journal of Applied Physics* **92**, 4414-4421 (2002).
14. G. Ghibaudo and T. Boutchacha, Electrical noise and RTS fluctuations in advanced CMOS devices, *Microelectron. Reliab.* **42**, 573-582 (2002).
15. P. Dutta and P. M. Horn, Low-frequency fluctuations in solids: $1/f$ noise, *Rev. Mod. Phys.* **53**, 497-516 (1981).
16. F. N. Hooge, T. G. M. Kleinpenning, and L. K. J. Vandamme, Experimental studies on $1/f$ noise, *Rep. Prog. Phys.* **44**, 479-531 (1981).
17. M. B. Weissman, $1/f$ noise and other slow, nonexponential kinetics in condensed matter, *Rev. Mod. Phys.* **60**, 537-571 (1988).
18. M. Surdin, Fluctuations in the thermionic current and the 'flicker effect', *J. Phys. Radium* **10**, 188-189 (1939).
19. F. N. Hooge, On the additivity of generation-recombination spectra. Part 2: $1/f$ noise, *Physica B* **336**, 236-251 (2003).
20. F. N. Hooge, $1/f$ noise is no surface effect, *Phys. Lett. A* **29a**, 139-140 (1969).
21. F. N. Hooge and L. K. J. Vandamme, Lattice scattering causes $1/f$ noise, *Phys. Lett. A* **66**, 315-316 (1978).
22. F. N. Hooge, Discussion of recent experiments on $1/f$ noise, *Physica* **60**, 130-144 (1972).
23. E. Simoen, and C. Claeys, On the flicker noise in submicron silicon MOSFETs, *Solid-State Electron.* **43**, 865-882 (1999).
24. C. Claeys, A. Mercha, and E. Simoen, Low-frequency noise assessment for deep submicrometer CMOS technology nodes, *J. Electrochem. Soc.* **151**, G307-G318 (2004).
25. L. K. J. Vandamme, X. Li, and D. Rigaud, $1/f$ noise in MOS devices, mobility or number fluctuations?, *IEEE Trans. Electron Devices* **41**, 1936-1945 (1994).

26. J. Chang, A. A. Abidi, and C. R. Viswanathan, Flicker noise in CMOS transistors from subthreshold to strong inversion at various temperatures, *IEEE Trans. Electron Devices* **41**, 1965-1971 (1994).
27. M. von Haartman, A.-C. Lindgren, P.-E. Hellström, B. G. Malm, S.-L. Zhang, and M. Östling, $1/f$ noise in Si and $Si_{0.7}Ge_{0.3}$ pMOSFETs, *IEEE Trans. Electron Devices* **50**, 2513-2519 (2003).
28. P. H. Handel, $1/f$ noise-an 'infrared' phenomenon, *Phys. Rev. Lett.* **34**, 1492-1494 (1975).
29. P. H. Handel, Fundamental quantum $1/f$ noise in semiconductor devices, *IEEE Trans. Electron Devices* **41**, 2023-2033 (1994).
30. Th. M. Nieuwenhuizen, D. Frenkel, and N. G. van Kampen, Objections to Handel's quantum theory of $1/f$ noise, *Phys. Rev. A* **35**, 2750-2753 (1987).
31. L. B. Kiss and P. Heszler, An exact proof of the invalidity of 'Handel's quantum $1/f$ noise model', based on quantum electrodynamics, *J. Phys. C: Solid State Phys.* **19**, L631-L633 (1986).
32. C. M. Van Vliet, A survey of results and future prospects on quantum $1/f$ noise and $1/f$ noise in general, *Solid-State Electron.* **34**, 1-21 (1991).
33. G. S. Kousik, C. M. Van Vliet, G. Bosman, and P. H. Handel, Quantum $1/f$ noise associated with ionized impurity scattering and electron-phonon scattering in condensed matter, *Advances in Physics* **34**, 663-702 (1985).
34. T. Musha and M. Tacano, Dynamics of energy partition among coupled harmonic oscillators in equilibrium, *Physica A* **346**, 339-346 (2005).
35. R. P. Jindal and A. van der Ziel, Phonon fluctuation model for flicker noise in elemental semiconductors, *J. Appl. Phys.* **52**, 2884-2888 (1981).
36. M. N. Mihaila, Phonon-induced $1/f$ noise in MOS transistors, *Fluctuation and Noise Letters* **4**, L329-L343 (2004).
37. S. V. Melkonyan, V. M. Aroutiounian, F. V. Gasparyan, and H. V. Asriyan, Phonon mechanism of mobility fluctuation equilibrium fluctuation and properties of $1/f$ noise, *Physica B* **382**, 65-70, (2006).

1. Fundamental noise mechanisms

PROBLEMS

1. A special electronic component is offered by five different manufacturers. The functionality of the component is the same for all designs, but the noise level differs. The resistance is 5 kΩ in all cases. A current $I = 1$ mA flows through the device in normal operation. The device has a bandwidth of 1 MHz and is used at room temperature $T = 300$ K. The devices have white noise only. The manufacturers give the following specifications:

A: There is only thermal noise
B: The noise temperature is 500 K
C: The RMS noise voltage is 15 µV.
D: The PSD of the noise voltage is 2×10^{-16} V²/Hz
E: The noise resistance is 10 kΩ

Arrange the components in order of increasing noise.

2. A semiconductor resistor ($R = 200$ Ω), biased at a constant current $I = 5$ mA shows $1/f$ noise with PSD $S_I(f) = 2.5 \times 10^{-19}/f$ A²/Hz. Calculate the root mean square of the noise current in the bandwidth 1 Hz – 10 kHz.

3. The circuit shown in Fig. 1-10 below is used for noise measurement of a diode. The current amplifier is assumed to have zero input impedance. Calculate the PSD of the noise current through the amplifier input.

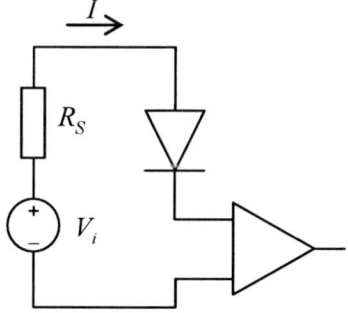

Figure 1-10. Circuit in example 3.

4. We study the noise of an n-type poly-silicon resistor with ohmic contacs. Fig. 1-11 shows the measured noise voltage PSD for the resistor. A DC current $I = 16.6$ µA flows through the resistor, which is held at room temperature $T = 300$ K.

The sample has the following dimensions: width = 10 µm, length = 100 µm, thickness = 1 µm. The doping concentration is 10^{17} cm^{-3}. What is the resistance of the sample?

Calculate the Hooge parameter α_H for the poly-Si resistor.

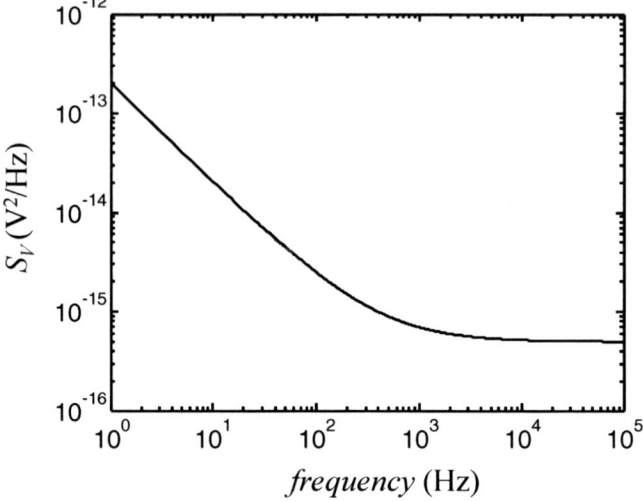

Figure 1-11. Measured noise voltage PSD vs. frequency in example 4.

Chapter 2

NOISE CHARACTERIZATION

1. INTRODUCTION

The measurement of noise is a complex task as the signal to be measured is very small (down to ~1 pA), usually in the presence of a much stronger DC bias current as well as undesired disturbances from electronic equipment. The measurement setup must be designed carefully with appropriate shielding and preferably using batteries as power sources to avoid disturbances to be injected in the circuits of interest. The measurements are usually performed in the frequency domain by measuring the power spectral density with a spectrum analyzer. If RTS noise is present, time domain analysis with the help of an oscilloscope is a valuable tool. A preamplifier is used to amplify the weak noise signal in order for it to be studied with the spectrum analyzer or oscilloscope. The typical setups and corresponding equipment used for low-frequency (LF) noise measurements are described in section 2. We also describe the analysis methods and give some practical advice. The low-frequency noise in a device is sensitive to the device technology, especially the presence of traps, defects and lattice damage. Therefore, important information about reliability and sensitive areas for the current transport can be obtained from noise studies.[1-3] Section 3 discusses the low-frequency noise measurements as a diagnostic tool.

2. LOW-FREQUENCY NOISE MEASUREMENTS

2.1 Measurement setup

The frequency range for low-frequency noise measurements is typically from 1 Hz to 1 MHz. Since the spectral shape also provides very important information, the power spectral density is the quantity that is preferably measured instead of, for example, noise power. The low-frequency noise measurement technique concerns sensing very weak signals, which makes it necessary to design the setup in order to minimize the internal noise as well as to prevent external disturbances from corrupting the measurement. By using batteries as power sources to bias the circuit, one can avoid disturbances from the power lines being injected into the circuit. On the other hand, filtered voltage sources may be perfectly good enough and allow better automation of the measurements (which can save many days in the lab). Shielding is important to prevent unavoidable disturbances from the outer world to interfere with the measurements. Electrical equipment connected to the power mains give rise to disturbances at 50 Hz or 60 Hz and at multiples thereof, and usually at other frequencies as well. Wireless units, mobile phones, radio transmitters etc provide disturbances in the MHz and GHz range. These signals are outside the bandwidth of the amplifier and the frequency range of interest for the measurements but the signals tend to mix and cause disturbance at lower frequencies also. Moreover, the pulse caused when a call is picked up or ended generates signal components especially at lower frequencies. Wireless and mobile phones are one of the biggest enemies in doing reliable noise characterizations and must be avoided in close proximity to the measurement setup.

Even if the disturbances can be reduced to an acceptable level, the measurement might still not be accurate. When doing the actual measurement, the total noise response at the output of the amplifier will be measured. The noise response from the other elements in the measurement setup should be (much) lower than the noise response from the device-under-test in order to make reliable measurements. It is therefore important to minimize the noise of the other elements in the setup, primarily the internal noise of the amplifier and the resistors in the bias circuit.

A typical low-frequency noise measurement configuration powered with batteries, which have been used for MOSFET noise characterization in the author's work, are presented in Fig. 2-1. On-wafer measurements are usually preferred for characterization of semiconductor devices; triax (or BNC) cables are used for that purpose to connect the device from the probe station to the bias circuit in the setup. The weak noise from the device is amplified by a low-noise amplifier and then fed to the spectrum analyzer which

2. Noise characterization

measures the power spectral density. The output from the amplifier is also monitored by an oscilloscope, which is important in order to detect presence of RTS noise and to make sure that the amplifier is not saturated. Two types of amplifiers are frequently used in low-frequency noise measurement setups. Fig. 2-1 describes a setup with a low-noise voltage amplifier (denoted by VA), which amplifies the voltage at its high impedance input to the output by a factor A. The setup can also be operated with a low-noise current amplifier (denoted by CA), which amplifies the current through its low-impedance input and delivers a voltage at the output amplified by the transimpedance gain G. An example of a typical low-frequency measurement setup that is fully automated is shown Fig. 2-2. This setup has programmable voltage sources and uses standard operational amplifiers for amplification. In the following subsections, we will discuss the design of the bias circuit and techniques for shielding the setup. The amplifiers and the spectrum analyzer are described in further detail in the section 2.2.

2.1.1 Shielding

Appropriate shielding should be used for the part of the setup where the desired signal is sensitive to disturbances. After amplification, the signal is large enough to become unaffected by the environment. Therefore, the bias circuit, the device-under-test and the amplifier should be shielded. The interfering electromagnetic signals, which consist of time dependent electric $E(t)$ and magnetic fields $H(t)$, can have a wide range of frequencies. Interference at 50/60 Hz and multiples thereof (harmonics) are especially troublesome.

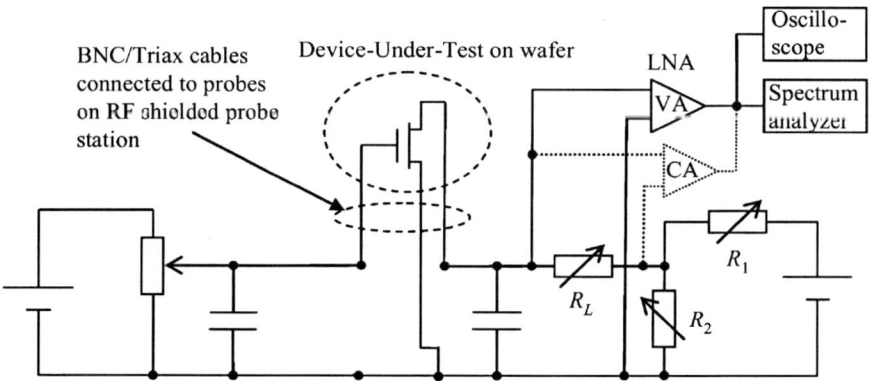

Figure 2-1. Low-frequency noise measurement setup employed with a voltage amplifier (VA) or a current amplifier (CA). When the current amplifier is used, R_L is disconnected.

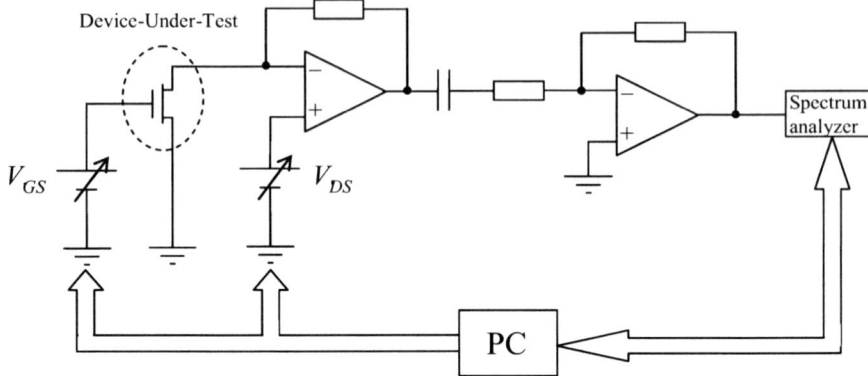

Figure 2-2. Automated low-frequency noise measurement setup (a redrawn and simplified version of the setup used by Chang and co-workers[4]).

When an electromagnetic wave propagates in a conductor, the amplitude is attenuated with distance from the conductor surface. The *skin depth* δ is the penetration depth where the amplitude of the wave has decreased by a factor e^{-1} and is given as

$$\delta = \frac{1}{\sqrt{\pi f \mu_0 \mu_r \sigma}}. \tag{2-1}$$

Here, f is the frequency of the wave, μ_0 the permeability of free space and μ_r the relative permeability (= 1 for most material, >> 1 for magnetic materials such as iron). Table 2-1 below lists the skin depth in some metals at various frequencies.

Table 2-1. Skin depths δ (mm) for various materials.[5]

Material	σ (S/m)	f = 50 Hz	1 kHz	1 MHz
Silver	6.17×10⁷	9.06 mm	2.03 mm	0.064 mm
Copper	5.80×10⁷	9.35	2.09	0.066
Aluminium	3.54×10⁷	11.96	2.67	0.085
Iron ($\mu_r \approx 10^3$)	1.00×10⁷	0.71	0.16	0.0050

Shielding the setup by a cage made of iron, a few mm thick, is sufficient and corresponds to several skin depths at 50 Hz as seen in Table 2-1.

2.1.2 Design of bias circuit

The bias circuit is used to power the device and set the currents and voltages to chosen values, the bias point. The requirements on the bias

2. Noise characterization

circuit is primarily to add as little noise as possible to the setup and allow flexible operation. As mentioned earlier, the circuit is preferably powered with batteries to aviod external disturbances. Since low-frequency noise measurements are very slow, completing noise meaurements at one bias point takes several minutes, automatic low-frequency noise measurement systems may save many hours in the lab. Such systems are easily constructed by using programmable voltage sources connected to a computer that controls the bias sweep and data collection. Metal film resistors should be used in the bias circuit since their LF noise is negligible, thus they only exhibit thermal noise.[6] The values of the resistors should be selected in such a way that the noise from the device-under-test is maximized at the output at the same time as the added thermal noise from the resistors is minimized. Thus, R_L in the setup in Fig. 2-1 should be larger than the channel resistance r_{ch} of the MOSFET. When the setup is used with the current amplifier, then $R_1//R_2$ should be small and R_L removed. The (large) shunting capacitor at the input is used for AC grounding the gate of the MOSFET. The output capacitor should be small, its purpose is to limit the bandwidth of the setup in order to reduce the impact of high frequency interfering signals.

2.2 Measurement equipment

2.2.1 Amplifiers

A low-noise amplifier (LNA) is used to amplify the weak noise signal before being monitored by the spectrum analyzer. However, the amplifier inevitably adds its own internal noise to the noise that we want to measure (from the device-under-test). Therefore, the internal noise of the amplifier sets the measurement limit of the system and must be minimized. The noise of the amplifier is modeled by two equivalent noise generators i_n and v_n at the input according to Fig. 2-3. Another commonly used measure of the amplifier noise performance is the noise factor F or the noise figure $NF = 10\log_{10}F$. The noise factor is defined in terms of the input and output signal-to-noise ratios (SNR = Signal power/Noise power) as

$$F = \frac{SNR_{in}}{SNR_{out}}. \qquad (2\text{-}2)$$

The noise factor can then be written as follows

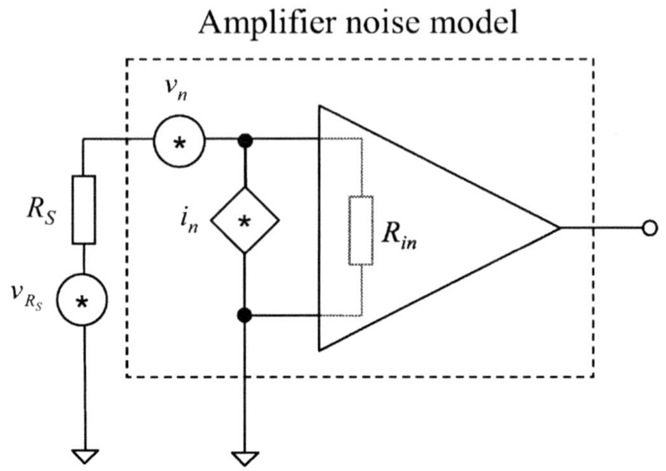

Amplifier noise model

Figure 2-3. Noise model for the amplifier with two equivalent input noise generators. R_{in} is the amplifier's input resistance.

$$F = \frac{\overline{v_{R_S}^2} + \overline{(v_n + i_n R_S)^2}}{\overline{v_{R_S}^2}} \quad (= \frac{\overline{v_{R_S}^2} + \overline{v_n^2} + \overline{i_n^2} R_S^2}{\overline{v_{R_S}^2}} \text{ if uncorrelated}). \quad (2\text{-}3)$$

Note that the noise factor depends on the source resistance R_S. The noise factor is equal to unity for an ideal noiseless amplifier, but $F > 1$ for any real amplifier. If several amplifiers are cascaded, the amplifier with lowest noise factor should be placed first (if the gain >> 1). This is understood from Friis formula[7] for cascaded amplifiers

$$F_{tot} = F_1 + (F_2 - 1)/A_{P,1} + (F_3 - 1)/A_{P,1}A_{P,2} + \ldots \quad (2\text{-}4)$$

where F_i and $A_{P,i}$ is the noise factor and available power gain for the i:th amplifier, respectively.

The requirements on a good low-noise amplifier to be used for sensitive noise measurements include properties such as ultra low internal noise, sufficient frequency range (DC to 100 kHz typically used in the measurements here), variable gain, and a wide dynamic range. A matched output (50 Ω) may also be desired.

Which type of amplifier should be selected for the noise characterization? As a general rule of thumb, the low-noise current amplifier outperforms the voltage amplifier at ultra-small currents where the input impedance of the device-under-test (R_{DUT}) is very large, for example a MOSFET biased in subthreshold. The voltage amplifier is often better to use at higher currents, such as a MOSFET biased in strong inversion. This is because the equivalent input current noise increases at lower gain settings (for the current amplifier) and the equivalent input noise voltage of the amplifier contributes when the impedance of the device is comparable or lower than the input impedance of the amplifier.

Let us analyse an example of a noise measurement with a current amplifier. A slightly simplified version of the setup in Fig. 2-1 is shown in Fig. 2-4(a). The device-under-test can for example be the channel of a MOSFET. The equivelent noise circuit is presented in Fig. 2-4(b), where the amplifier is represented with two equivalent input noise generators as in Fig. 2-3. The resistor and the device are replaced with Norton equivalents containing their corresponding noise generators in parallel with noiseless resistances.

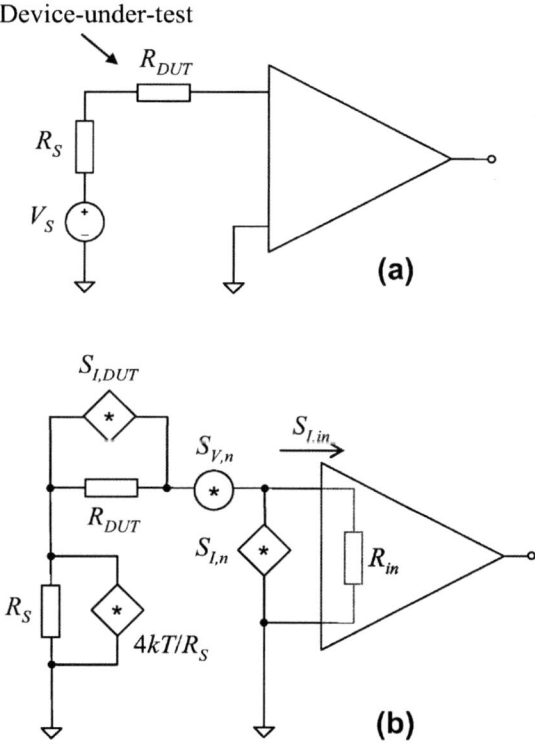

Figure 2-4. (a) Simple noise measurement setup with a current amplifier. (b) Noise equivalent circuit of the setup in (a).

Using the superposition principle, the PSD of the input noise current to the amplifier can be written as

$$S_{I,in} = \frac{4kT}{R_S} \frac{R_S^2}{(R_S + R_{DUT} + R_{in})^2} + S_{I,DUT} \frac{R_{DUT}^2}{(R_S + R_{DUT} + R_{in})^2}$$
$$+ S_{V,n} \frac{1}{(R_S + R_{DUT} + R_{in})^2} + S_{I,n} \frac{(R_S + R_{DUT})^2}{(R_S + R_{DUT} + R_{in})^2}.$$
(2-5)

It is evident that $R_S = 0$ eliminates the thermal noise contribution from R_S (first term), which is advantageous for a more sensitive measurement of $S_{I,DUT}$. For large values of R_{DUT}, $S_{I,n} \cdot (R_{DUT})^2 \gg S_{V,n}$ and reliable noise measurements of the device can be made when $S_{I,DUT} > S_{I,n}$. If $S_{I,DUT} \approx S_{I,n}$, then $S_{I,n}$ must be well characterized (not just taken from data sheets) in order to accurately extract $S_{I,DUT}$. Note that the voltage noise source $S_{V,n}$ contributes appreciably when R_{DUT} is small.

2.2.2 Spectrum analyzer

A spectrum analyzer is used to measure and analyze a signal in the frequency domain. Modern spectrum analyzers utilize the discrete Fast Fourier Transform (FFT) algorithm to convert the measured signal from the time domain to the frequency domain. A simplified block diagram describing the function of a FFT spectrum analyzer is shown in Fig. 2-5. After initial attenuation/amplification and low-pass filtering, the signal is sampled and digitized. Digital signal processing is then performed for data manipulation and implementation of the FFT algorithm.

In order to get a more intuitive view of what the spectrum analyzer is doing, one can think of how a swept-tuned analyzer operates. This type of analyzer use a bandpass filter to study the signal in a small frequency interval Δf centered at frequencies f_1, f_2, etc. The signal power is measured after the bandpass filtering and divided by Δf in order to achieve $S(f_1)$, $S(f_2)$, etc.

When making low signal level measurements, there are some important factors and analyzer settings to consider. The noise floor of the analyzer limits how small signals that can be measured with the spectrum analyzer. The sensitivity of the spectrum analyzer should be selected as high as possible (minimizing the attenuation) without overloading the input. By using a preamplifier, as discussed previously, the measurement sensitivity is greatly improved.

2. Noise characterization

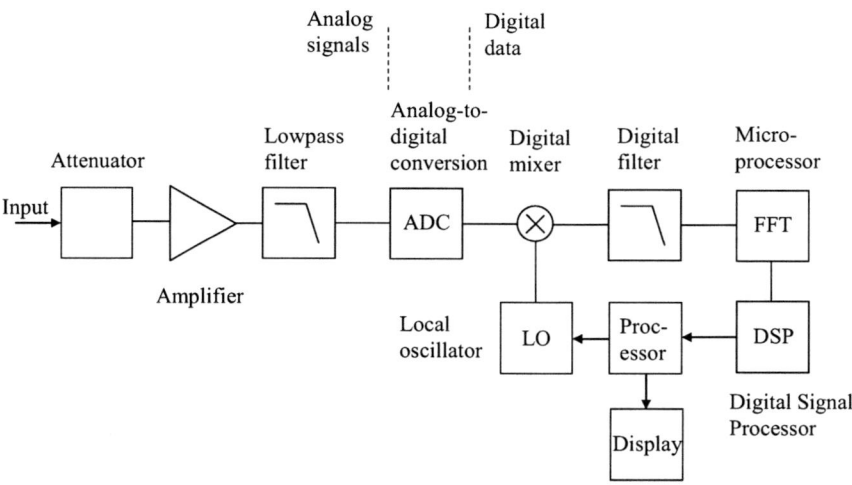

Figure 2-5. Block diagram of a FFT spectrum analyzer (source: Agilent[8]).

The frequency resolution of the spectrum measurement is determined by the *resolution bandwidth*, *frequency span* and number of frequency points. The display resolution, which is equal to the frequency span divided by the number of frequency points minus one, is improved for narrower spans and more frequency points. The frequency resolution is ultimately limited by the resolution bandwidth. A narrower resolution bandwidth not only improves frequency resolution, but also lowers the noise floor of the measurement because there is less noise power in a narrower bandwidth (as discussed in the previous chapter). However, one drawback by reducing the resolution bandwidth is that it makes the measurements slower. Normally, the resolution bandwidth is adjusted automatically when a different span is selected.

The FFT analysis assumes that the signal is periodic from time record to time record. However, in reality this is not the case which causes a broadening of the signal energy over frequency during the FFT operation. To circumvent the spectral broadening, the signal is multiplied with a time domain weighting function, called *window function*, to make the signal periodic in the time record. Most spectrum analyzers have several choices of window functions, the most common types are the Hanning, rectangular, Gaussian top and flattop windows. Each window function has different advantages and disadvantages; there is a trade-off between frequency and amplitude resolution. The Hanning window provides a good frequency resolution; it can be used for general purpose and is for the most cases a good choice for LF noise measurements. The flattop window gives high amplitude accuracy at the expense of lower frequency resolution.

The analyzer presents (among other options) the power spectral density of the voltage noise at the analyzer input in units of V^2/Hz. Averaging must be used for reliable measurements of noise, 50-100 averages is typically enough. A good idea when performing the measurements is to use a narrow frequency span for good frequency resolution and combine measurements from several spans to make up the total spectrum.

2.3 Frequency and time domain analysis

The standard way to analyze noise, which gives most information about its properties, is to make the measurements with a spectrum analyzer and study the noise in the frequency domain. For RTS noise, time domain measurements with an oscilloscope can give supplemental or even sufficient information.

Noise spectral data are typically collected for a number of transistor bias conditions and then transferred to a PC for further processing and analysis. The power spectral density at the amplifier input is obtained by dividing with the gain squared. Fig. 2-6 shows two noise spectra, one (a) with pure $1/f$ noise and one (b) with superimposed $1/f$ noise and g-r noise. The $1/f$ noise is modeled according to Eq. (1-33) with a magnitude and a frequency exponent. It is common that the LF noise is extracted at a selected frequency, for example 10 Hz, in order to study the noise variation with bias and between different devices. If the measured noise spectrum contains g-r noise, $1/f^\gamma$ noise, and white noise components, it is often a good idea to separate these noise mechanisms as shown in Fig. 2-6(b) and then extract relevant parameters.

Time domain measurements are important in order to characterize RTS noise and can for example be performed with an oscilloscope. The RTS noise pulse heights and time durations in the upper and lower RTS level are collected from the oscilloscope data in order to analyze the RTS noise. Pulse trains with 20-100 transitions should be recorded for each bias point for reliable extraction of the mean time durations; an uninterrupted stream containing all the necessary transitions is ideally preferred. An example of a time domain noise measurement showing random noise pulses is shown in Fig. 2-7.

2. Noise characterization

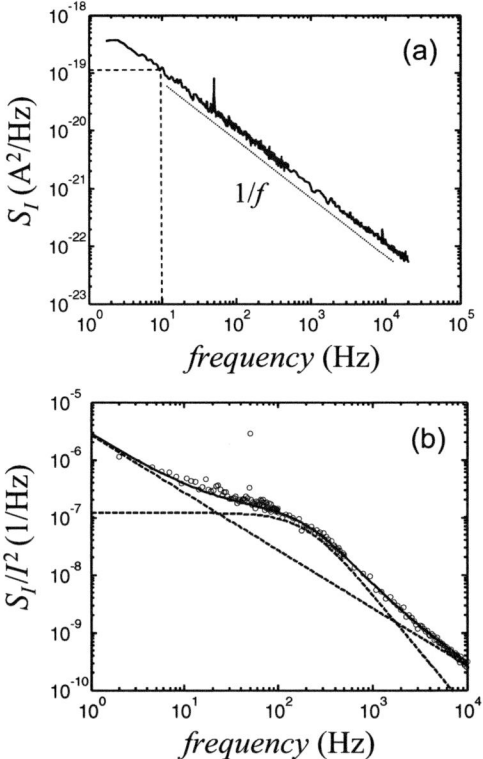

Figure 2-6. (a) Noise spectrum showing pure $1/f$ noise. (b) A more complex noise spectrum, which can be decomposed into a g-r noise and a $1/f$ noise component. Note the peaks at 50 Hz in the spectra due to external disturbances.

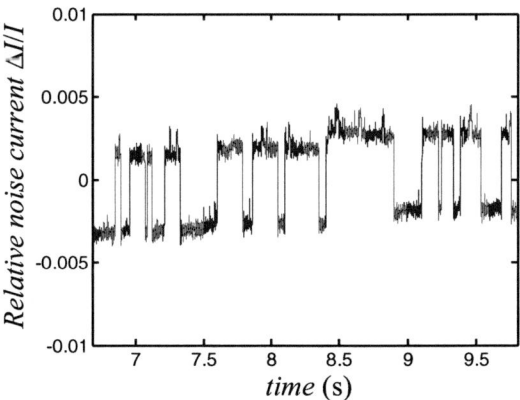

Figure 2-7. Random-telegraph-signal (RTS) noise.

The RTS noise process is characterized by the mean values of the pulse height and the time durations (time constants). The pulse heights follow the normal distribution in most cases, as exemplified in Fig. 2-8(a). The probability for switching between the RTS levels, on the other hand, is Poisson distributed. The occurrence of time durations between t and $t + \Delta t$ is then given as[9]

$$\text{Counts between } t \text{ and } t + \Delta t \propto \exp(-t/\tau)/\tau. \qquad (2\text{-}6)$$

The mean value of the time durations τ for the RTS process can then be computed from a semi-logarithmic histogram analysis of counts versus time, as shown in Fig. 2-8(b). Note that the mean value of the Poisson process is not equal to the arithmetic mean. Therefore, a histogram analysis should be performed to first validate that the RTS process is Poisson distributed and then the time constants are calculated according to Eq. (2-6).

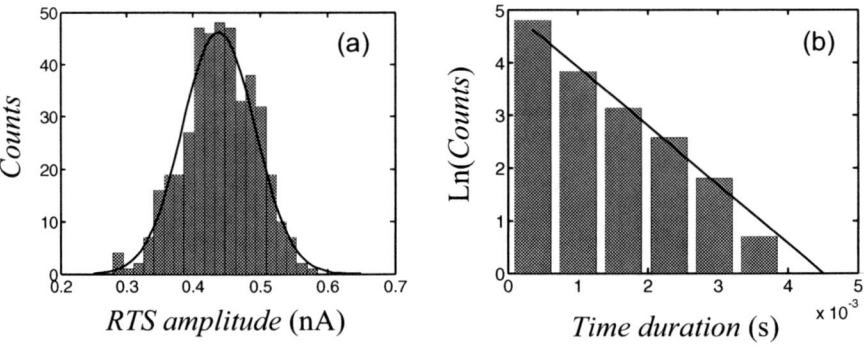

Figure 2-8. (a) Distribution of RTS amplitudes (pulse heights) resembling a normal distribution. (b) Semi-logarithmic histogram plot of the time durations. The line is the best linear fit to the data; the inverse slope corresponds to the time constant.

The RTS noise analysis becomes more complicated if there are several traps involved that generate RTS noise so that there are more than two RTS levels. For an accurate analysis, the pulse heights and time constants associated with different traps should be extracted separately. If the pulse heights and/or time constants differ enough, two (or more) peaks can be observed in the histograms which allow the mean values to be calculated for each pulse type. Once the mean values (ΔI, τ_l and τ_h) are calculated for each pulse type, the noise spectrum can be reconstructed by summing the spectral contributions for each pulse type calculated from Eq. (1-32). Note that one cannot study ΔI, τ_l and τ_h separately just from the measured PSD. Therefore, the time domain measurements are needed to obtain detailed information about the RTS noise. The PSD should, however, be measured also if RTS noise is present in order to study other noise mechanisms than the RTS noise.

2.4 Some practical advice

Performing reliable LF noise measurements require some knowledge and experience due to the difficulty of measuring the weak noise signals. Even though commercially available automated noise measurement systems have appeared in the last years that facilitate the measurement task, a great deal of hands-on work is still involved. Below is a list of some helpful practical advice in order to perform LF noise measurements:

- Shielding is important in order to avoid disturbances. Enclose the bias circuitry, the device-under-test and amplifier in an iron cage (spectrum analyzer and oscilloscope outside).
- Turn off unused equipment that might disturb the measurements, especially wireless and mobile phones should not be used in the vicinity of the measurement setup. Switching of lights and equipments introduce disturbances especially at low frequencies.
- Use short well-shielded triax or coax cables for connections. Avoid any open connectors in the cage.
- Use metal film resistors and batteries in the bias circuitry for lowest noise. Make sure that the batteries provide a stable bias current.
- Use a preamplifier with low internal noise.
- The noise at the output can stem from many different sources. Calculate the expected noise level at the output from the different sources (device-under-test, resistors, amplifier etc). Measure the white noise and compare with the expected thermal noise or shot noise level. Use the result for calibration of your setup (or your calculations).
- Bad contacts, for example due to worn-out probes or too gentle probing force can introduce noise that exceeds the noise from the device-under-test. A good idea is to check that the probe pressure does not affect the measurement.
- Use narrow frequency spans for good frequency resolution when measuring the spectral density. Disturbances can easily be identified as narrow peaks at certain frequencies when the resolution is good enough.
- Be patient! Low-frequency noise measurements are very slow since the measurements typically are made down to a few Hz. Still, if "random" disturbances (ringing phones, colleagues switching on and off equipment

etc) occur, monitoring the measurements is advantageous since any corrupted measurements can be remade.
- It is often better to perform LF noise measurements on quieter evenings and weekends (harm on social life unaccounted for).
- Parasitic capacitances in the measurement setup may lead to a limited bandwidth. If the white noise level drops below the expected one at high frequencies (usually above ~10^4 Hz), this might be the reason.
- Coherence measurements (given from cross-correlation of two simultaneously measured noise signals) can be useful to find the dominant noise source in transistors, for example if the noise in the gate current dominates the LF noise in a MOSFET.
- It is better to perform measurements on small devices since they generate more noise (easier to measure). However, be aware of larger device-to-device variations (not least in terms of noise) for small devices. Characterizing MOSFETs with a 10 μm^2 gate area is a good compromise.

2.5 Other noise measurement methods

In the measurement setup for LF noise characterization of MOSFETs, we implicitly assumed that the source resistance connected to the gate is unimportant. For a complete characterization of the equivalent input noise sources v_n and i_n of an amplifier, at least two measurements are necessary. By short-circuiting the input of the amplifier, only v_n contributes (see Fig. 2-3) and can thus be determined. The current noise generator i_n can be characterized with an open input (very large $R_S \gg R_{in}$) since the v_n contribution then is zero. If v_n and i_n are correlated (often they are not), a third measurement with an appropriate R_S is necessary in order to determine the correlation. The reason why this procedure is not used in the noise measurements of the MOSFET is that the noise current generator at the input of the MOSFET is very small and can be neglected for any normal value of R_S. However, the gate leakage current can be considerable for very thin gate dielectrics. In such case, the noise in gate current can give rise to a significant noise contribution at the output.

Although LF noise measurements are our primary concern, we are going to briefly mention high-frequency noise characterization methods. At high frequencies (MHz to GHz range), only white noise sources need to be considered. On the other hand, the MOS capacitances and parasitic inductances must be included in the noise model. At high frequencies there is a pronounced effect of thermal noise in the channel that couples to the gate, so called *induced gate noise*. The actual characterization of the noise is a different science than the LF noise measurements. At microwave frequencies, the signals must be treated as travelling waves. When a wave is incident on a boundary between two different media (different impedances), a portion of the incident wave power is transmitted and a portion is reflected. A typical high frequency noise measurement setup contains a known noise source, a source impedance tuner and a noise figure meter. The noise figure is typically measured for a range of source reflection factors (covering as much as possible of the impedance plane) in order to find the minimum noise figure (NF_{min}). From this measurement two other parameters are also obtained: the optimum source impedance $Z_{S,opt}$ (real and imaginary parts) and the equivalent noise resistance r_n. The noise factor is a parabolic function of the source admittance $Y_S = (Z_S)^{-1}$ according to[10]

$$F = F_{min} + \frac{r_n}{\text{Re}(Y_S)}|Y_S - Y_{S,opt}|^2 \qquad (2\text{-}7)$$

3. NOISE AS A DIAGNOSTIC TOOL

Low-frequency noise measurements can be used as a valuable tool for quality and reliability evaluations of electronic devices. Slow (deep) traps in the gate oxide of a MOSFET, situated close to the quasi-Fermi level energy, can for example be probed by LF noise measurements. LF noise measurements are therefore a good complement to charge-pumping measurements which primarily are used to characterize surface states in the middle of the band gap. LF noise characterizations can also give important insights about other types of traps and defects and their kinetics, electron scattering especially with phonons and lattice defects. The LF noise analysis can further help identify sensitive areas for current transport and determine the impact of technology on device quality and reliability. This chapter describes the LF noise characterization technique as a diagnostic tool for the abovementioned evaluations in MOSFETs. A reader that is not familiar with MOSFET noise sources and $1/f$ noise mechanisms is advised to read chapter 3 first.

3.1 Determining the dominant noise source

When an unknown device is characterized, the first step is to determine the dominant noise mechanism and the spatial location of the noise source. This is accomplished by studying the LF noise for different bias conditions, device parameters and technological factors. The location of the dominant noise source is revealed by analyzing the bias and geometry dependence of the noise. LF noise characterizations for diagnostic purposes are preferably performed at small drain voltages (often around 50 mV). The LF noise in a

MOSFET is generated in the channel and the extrinsic S/D access series resistances. Fig. 2-9 shows a cross-sectional transmission electron microscopy (TEM) image of a MOSFET where these two noise sources are indicated. The channel noise usually dominates at low and medium current levels, but the noise from the S/D resistances can make a significant contribution at large currents. The measured drain current noise can be written as (see further next chapter)

$$S_{I_{D,tot}} = \frac{S_{I_{D,ch}} + g_{ch}^2 R_{SD}^2 S_{I_{RSD}}}{[1 + g_{ch}(R_{SD} + R_L)]^2} \qquad (2-8)$$

$S_{I_{D,tot}}$ – measured drain current noise (PSD) at output

$S_{I_{D,ch}}$ – drain current noise (PSD) in channel

$S_{I_{RSD}}$ – drain current noise (PSD) in S/D resistance

R_{SD} – total S/D resistance. $R_S = R_D = R_{SD}/2$

g_{ch} – channel conductance

R_L – load resistance between source and drain.

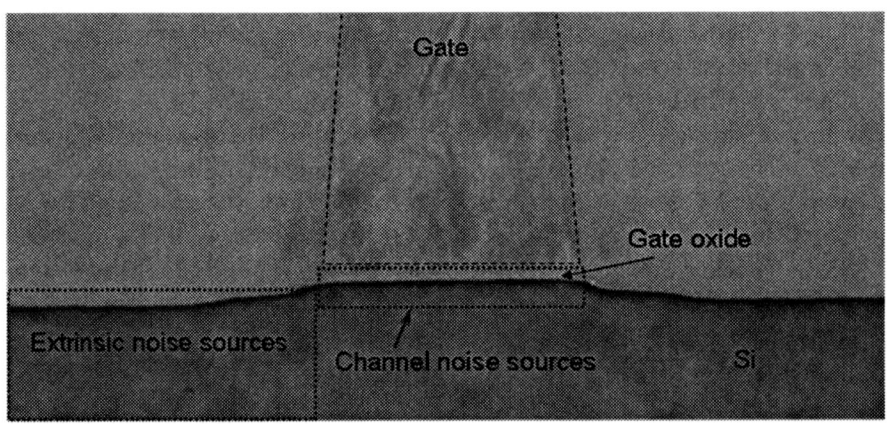

Figure 2-9. Cross-sectional TEM image of a MOSFET. The two most important $1/f$ noise sources are indicated.

The channel drain current noise depends on drain current, gate length and the gate voltage, whereas the noise from the S/D resistance is independent of the latter two variables

2. Noise characterization

$$S_{I_{D,ch}} \propto \frac{I_D^2}{fWL(V_{GS}-V_T)^p} \quad \text{(in strong inversion)} \tag{2-9}$$

$$S_{I_{R_{SD}}} \propto \frac{I_D^2}{f}. \tag{2-10}$$

Here p is an exponent that depends on the mechanism for $1/f$ noise in the channel. V_{GS} is the gate-source voltage, V_T the threshold voltage, I_D the drain current, W the gate width and L the gate length. If the noise stemming from the S/D resistance is dominant, the drain current noise at constant drain current should be independent of the gate length, as seen from Eqs. (2-8) to (2-10). On the other hand, the noise originating from the channel increases with decreasing gate length. The drain current noise divided with the drain current squared, called the normalized drain current noise, decreases with increasing $I_D \propto (V_{GS} - V_T)$ as follows from Eq. (2-9). If the noise in the S/D resistance dominates, the measured normalized drain current noise is instead expected to increase with the drain current squared since $g_{ch} \propto I_D$ in the linear regime (see Eq. 2-8). Fig. 2-10 shows a simulation of the normalized drain current noise using Eqs. (2-8) to (2-10) for MOSFETs with various gate lengths, which describes the different situations above. In this example, I_D is varied by varying V_{GS} at a constant V_{DS}.

Having determined the dominant noise source, remedies to reduce the noise can be sought out. In this process, information about the dominant noise mechanism is highly desired. The noise mechanism can be revealed by studying the bias dependence of the low-frequency noise. For a MOSFET the gate voltage is typically varied, which modifies the inversion carrier density. The dominant source of the $1/f$ noise, mobility fluctuation or number fluctuation noise, can then be identified in strong inversion by analyzing the resemblance with Eq. (2-9) and extraction of the exponent p. In practice, this is not straightforward. Both number and mobility noise may contribute to the measured noise with similar magnitudes and their relative strength change with bias. The drain current noise should be characterized over a wide bias range from subthreshold to the strong inversion region in order to get as much information as possible. With that said, one should be careful by determining the noise mechanism only from the bias dependence.

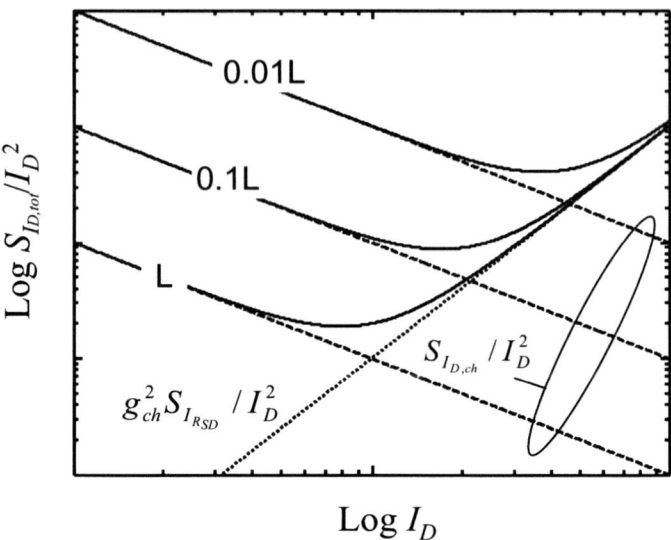

Figure 2-10. Simulation of the normalized drain current noise at the S/D output of a MOSFET using Eqs. (2-8) to (2-10). $p = 1$ was used in Eq. (2-9). I_D is varied by varying V_{GS}.

From theory, the Hooge mobility fluctuations is expected to show $p = 1$, whereas number fluctuations should result in a p around 2 in strong inversion.[11] If number fluctuations prevail, traps in the gate oxide are likely responsible for the LF noise. In case of mobility fluctuation noise, the noise is generated in carrier scattering processes, mainly the phonon scattering. However, there are usually deviations from the simple theory. The trap density may vary with energy, mobility fluctuations correlated to the fluctuations in the number of carriers contribute in strong inversion (the gate voltage dependence is roughly the same as for the Hooge noise), and the Hooge parameter can depend on the electric field to mention some complicating effects. We will give a more detailed description of the 1/f noise in MOSFETs and further discuss these complicating effects in the next chapter. It is often necessary to measure the LF noise over several decades of drain currents at a constant drain-source voltage. Investigating the substrate bias dependence of the low-frequency noise can provide additional information, as described in chapter 3. Correlating the noise level to other device parameters such as oxide charge density, interface state density, carrier mobility (especially phonon or Coulomb scattering limited mobility), oxide thickness (if varied in the experiments), etc, can help to establish the noise origin.

The information obtained about the location of the noise sources and the dominant noise mechanisms provides an understanding of the underlying physics and the possible measures that can be taken to improve the noise performance. The trap density and Hooge parameter can be used as figures-of-merit for a given technology or material system. In chapter 4, the 1/ƒ noise results derived for a wide range of MOS technologies are summarized. If traps in the gate oxide are found to govern the 1/ƒ noise, reducing the trap density by an improved gate oxidation process will reduce the noise.[12] If mobility fluctuations prevail, improved crystalline quality in the channel and reduced surface roughness can result in lower 1/ƒ noise.[1,11,13] A strained channel could also be beneficial. For both mechanisms, the utilization of a buried channel potentially gives improved noise performance.[14-16] The level of noise from the S/D regions can be lowered by decreasing the S/D resistance, avoiding current crowding and improving the quality of the contacts.[1]

3.2 Characterization of traps by using RTS noise

G-r noise, RTS noise and number fluctuation 1/ƒ noise in the drain current of a MOSFET originate from traps in the gate oxide. G-r noise and RTS noise can stem from traps at other locations also, for example the depletion region in the substrate, but it is rather rare. G-r and RTS noise are only important close to the quasi-Fermi level energy, and are therefore very bias and temperature sensitive. For RTS noise, only one trap is active, while g-r noise can be generated from one or several traps with equal time constants. RTS noise is therefore only observed in small devices or/and devices with a low background noise. The total noise can be decomposed in a g-r and a $1/f^\gamma$ noise component when RTS noise is present in the time domain. RTS noise can be observed on top of the mobility 1/ƒ noise in MOSFETs with small gate area (usually below 1 µm^2) if the following criterion on the number of carriers in the channel is fulfilled[17]

$$N < 1/4\pi\alpha_H \tag{2-11}$$

where α_H is the Hooge parameter for the 1/ƒ noise. Obviously, the occurrence of RTS noise falls off with increasing N (increasing gate voltage overdrive). If the $1/f^\gamma$ noise and the RTS noise have the same origin, traps in the gate oxide, the occurrence of RTS noise depends on gate area but not bias (except if the trap density is bias dependent). The number of traps that can generate $1/f^\gamma$ noise can be estimated according to

$$\text{Number of traps} = 4kTWLN_t z \tag{2-12}$$

where z is the tunneling distance of a carrier from the gate oxide/channel interface, maximum ~3 nm, and $4kT$ is the energy around the quasi-Fermi level where the traps are distributed. RTS noise can be observed if the number of available traps is small. The relative drain current amplitude is related to the trap position z_t ($z_t = 0$ at the gate oxide/channel interface) according to[18,19]

$$\frac{\Delta I_D}{I_D} = \frac{q}{WL}\left[\frac{g_m}{I_D}\frac{1-z_t/t_{ox}}{C_{ox}} + \alpha\mu_{eff}\right] \qquad (2\text{-}13)$$

where t_{ox} is the gate oxide thickness, C_{ox} the oxide capacitance, g_m the transconductance and α a scattering coefficient (in Vs/C). Note that the expression is not valid in the subthreshold region (see further chapter 3). The trap depth can also be extracted from the variation of the emission time with gate voltage[3]

$$d\ln(\tau_e)/dV_G \approx qz_t/(t_{ox}kT). \qquad (2\text{-}14)$$

The trap position along the channel can be estimated from the variation of τ_c/τ_e with drain voltage.[3] The capture and emission times, τ_c and τ_e, are in general governed by Shockley-Read-Hall statistics[20]

$$\tau_c = \frac{1}{\sigma_e n_s v_{th}} \quad \text{and} \quad \tau_e = \frac{\tau_c}{g \cdot e^{(E_T-E_F)/kT}} \qquad (2\text{-}15)$$

where n_s is the surface carrier concentration, $\sigma_{e(h)}$ is the electron (hole) capture cross section, and g is the degeneracy factor usually taken as unity for electrons. It is usually observed that τ_c varies inversely with the gate voltage overdrive but is weakly dependent on temperature. τ_e, on the other hand, decreases exponentially with temperature but is approximately constant with gate bias.

3.3   Characterization of traps by using 1/f noise

While the energy level and spatial location of a single trap can be determined from analysis of the RTS noise, the distribution of traps versus energy and oxide depth is characterized from the frequency and bias dependence of the number fluctuation $1/f^\gamma$ noise.[21] The trap density can be evaluated by (see next chapter)

2. Noise characterization

$$N_t = \frac{fWLC_{ox}^2 S_{V_G}}{q^2 kT\lambda \left(1 + \alpha\mu_{eff} C_{ox} I_D / g_m\right)^2} \quad [\text{cm}^{-3}\text{eV}^{-1}] \qquad (2\text{-}16)$$

$S_{V_G} = S_{I_D}/g_m^2$, equivalent input gate voltage noise

λ – tunneling attenuation length in gate oxide (≈ 1 Å for Si/SiO$_2$).

N_t is the density of traps at the quasi-Fermi level and can be found as a function of energy if the gate voltage noise is measured as a function of gate voltage. The surface quasi-Fermi level is found from solving the equations below (note: no analytic solution for ψ_s from Eq. 2-17)

$$V_G = V_{fb} + \psi_s + \frac{\sqrt{2q\varepsilon_{Si} N_a}}{C_{ox}} \sqrt{\psi_s + \frac{kT}{q} e^{q(\psi_s - 2\psi_B)/kT}} \qquad (2\text{-}17)$$

$$\psi_B = \frac{kT}{q} \ln\left(\frac{N_a}{n_i}\right) \qquad (2\text{-}18)$$

$$E_C - E_{F,n} = 0.56 + q\psi_B - q\psi_s \quad \text{(for nMOS)}. \qquad (2\text{-}19)$$

ψ_s is the surface potential, ψ_B the difference between the Fermi level and the intrinsic level, N_a the doping concentration, E_C the energy at the conduction band edge, $E_{F,n}$ the quasi-Fermi energy level and ε_{Si} the silicon permittivity. However, one must be cautious with this kind of analysis, the bias dependence of the LF noise could stem from a completely different mechanism. The interpretation that the LF noise behaviour is explained by a an energy dependent trap density is supported by some groups,[22,23] but is generally not accepted.

The gate voltage noise spectrum can also be used to estimate the depth dependence of the trap density. The depth is calculated from the frequency according to

$$z = \lambda \cdot \ln(1/2\pi f \tau_0) \qquad (2\text{-}20)$$

Here the time constant τ_0 is usually taken as 10^{-10} s. Fig. 2-11 illustrates the trap density profile for a pMOSFET with 5-nm ALD Al$_2$O$_3$ as gate dielectrics. The low-frequency noise was measured between 1 and 20 kHz.

An interfacial oxide, around 1-nm thick, was found to be present between the Al_2O_3 and the channel. As seen in Fig. 2-11, the trap density is higher in the bulk of the Al_2O_3 gate dielectrics than close to the SiO_2/Al_2O_3 interface. This exercise is an example of the usefulness of low-frequency noise measurements to characterize slow oxide traps. This type of trap is difficult to analyze with other methods. Standard charge-pumping techniques probe traps in the middle of the bandgap and situated very close to the oxide/channel interface. Therefore, low-frequency noise measurements fill an important purpose in the device evaluation process. Specialized charge-pumping techniques can also give information about the trap distribution versus depth. The values for the trap density in high-k gate dielectrics extracted from such techniques are in the same range as those obtained by noise measurements.[24-26]

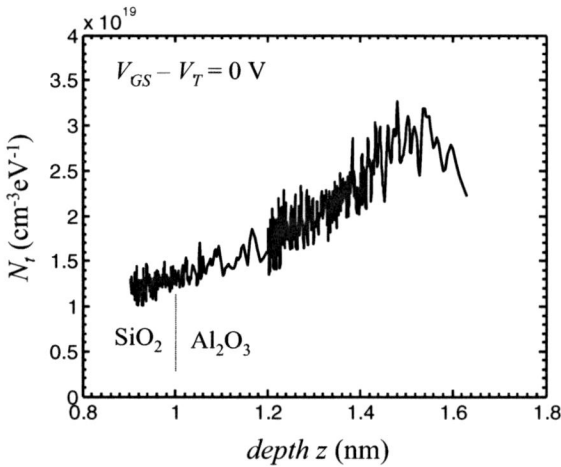

Figure 2-11. Oxide trap density vs. depth in the gate oxide ($z = 0$ at the oxide/channel interface).

SUMMARY

- A typical LF noise measurement setup consists of a bias circuit, a preamplifier and a spectrum analyzer. An oscilloscope can also be useful, especially in order to detect and study RTS noise.

- The noise signal that one wants to measure is usually very small; the measurements can therefore easily be affected by external disturbances and noise generated from other elements than from the device-under-test.
- Appropriate shielding around the setup is important to prevent disturbances to interfere with the noise signal that one wants to measure.
- An amplifier adds its own internal noise when it amplifies a signal from the input to the output. The noise of the amplifier can be modeled with two equivalent noise sources at the input.
- An amplifier with low noise is important for sensitive noise measurement since the noise of the amplifier sets the measurement limit.
- LF noise measurements can be used as a diagnostic tool to obtain information about device quality and reliability and study the impact of technology on these properties.
- The $1/f$ noise and g-r noise (studied in the frequency domain) and the RTS noise (studied in the time domain) can be used to study traps. $1/f$ noise can also potentially provide insights about lattice damage, electron scattering processes and sensitive regions for current transport.
- The bias and geometry dependence of the LF noise can be used to determine the mechanism and location of the dominant LF noise source.

REFERENCES

1. L. K. J. Vandamme, Noise as a diagnostic tool for quality and reliability of electronic devices, *IEEE Trans. Electron Devices* **41**, 2176-2187 (1994).
2. C. Claeys and E. Simoen, Impact of advanced processing modules on the low-frequency noise performance of deep-submicron CMOS technologies, *Microelectron. Reliab.* **40**, 1815-1821 (2000).
3. G. Ghibaudo and T. Boutchacha, Electrical noise and RTS fluctuations in advanced CMOS devices, *Microelectron. Reliab.* **42**, 573-582 (2002).
4. J. Chang, A. A. Abidi, and C. R. Viswanathan, Flicker noise in CMOS transistors from subthreshold to strong inversion at various temperatures, *IEEE Trans. Electron Devices* **41**, 1965-1971 (1994).
5. D. K. Cheng, *Field and wave electromagnetics* (Addison-Wesley, Reading, 1989).
6. C. D. Motchenbacher and J. A. Connelly, *Low-noise electronic system design* (John Wiley & Sons, New York, 1993).
7. H. T. Friis, Noise figures of radio receivers, *Proc. Institute of Radio Engineers (IRE)* **32**, 419-422 (1944).
8. Agilent Technologies 89410A/89441A Operator's guide (Hewlett-Packard, Washington, 2000).
9. M. J. Kirton and M. J. Uren, Noise in solid-state microstructures: a new perspective on individual defects, interface states and low-frequency ($1/f$) noise, *Advances in Physics* **38**, 367-468 (1989).
10. C. H. Chen and M. J. Deen, High frequency noise of MOSFETs I Modeling, *Solid-State Electron.* **42**, 2069-2081 (1998).

11. L. K. J. Vandamme, X. Li, and D. Rigaud, $1/f$ noise in MOS devices, mobility or number fluctuations?, *IEEE Trans. Electron Devices* **41**, 1936-1945 (1994).
12. E. Simoen, and C. Claeys, On the flicker noise in submicron silicon MOSFETs, *Solid-State Electron.* **43**, 865-882 (1999).
13. P. Gaubert, A. Teramoto, T. Hamada, M. Yamamoto, K. Kotani, and T. Ohmi, $1/f$ noise suppression of pMOSFETs fabricated on Si(110) and Si(100) using an alkali-free cleaning process, *IEEE Trans. Electron Devices* **53**, 851-856, 2006.
14. M. von Haartman, A.-C. Lindgren, P.-E. Hellström, B. G. Malm, S.-L. Zhang, and M. Östling, $1/f$ noise in Si and $Si_{0.7}Ge_{0.3}$ pMOSFETs, *IEEE Trans. Electron Devices* **50**, 2513-2519 (2003).
15. L. K. J. Vandamme, Bulk and surface $1/f$ noise, *IEEE Trans. Electron Devices* **36**, 987-992 (1989).
16. B. Cretu, M. Fadlallah, G. Ghibaudo, J. Jomaah, F. Balestra, and G. Guégan, Thorough characterization of deep-submicron surface and buried channel pMOSFETs, *Solid-State Electron.* **46**, 971-975 (2002).
17. T. G. M. Kleinpenning, On $1/f$ noise and random telegraph noise in very small electronic devices, *Physica B* **164**, 331-334 (1990).
18. O. Roux Dit Buisson, G. Ghibaudo, and J. Brini, Model for drain current RTS amplitude in small-area MOS transistors, *Solid-State Electron.* **35**, 1273-1276 (1992).
19. N. V. Amarasinghe, Z. Çelik-Butler, and A. Keshavarz, Extraction of oxide trap properties using temperature dependence of random telegraph signals in submicron metal-oxide-semiconductor field-effect transistors, *J. Appl. Phys.* **89**, 5526-5532 (2001).
20. W. Shockley and W. T. Read, Jr., Statistics of the recombinations of holes and electrons, *Phys. Rev.* **87**, 835-842 (1952).
21. R. Jayaraman and C. G. Sodini, A $1/f$ noise technique to extract the oxide trap density near the conduction band edge of silicon, *IEEE Trans. Electron Devices* **36**, 1773-1782 (1989).
22. J. H. Scofield, N. Borland, and D. M. Fleetwood, Reconciliation of different gate-voltage dependencies of $1/f$ noise in n-MOS and p-MOS transistors, *IEEE Trans. Electron Devices* **41**, 1946-1952 (1994).
23. M. J. Prest, A. R. Bacon, D. J. F. Fulgoni, T. J. Grasby, E. H. C. Parker, T. E. Whall, and A. M. Waite, Low-frequency noise mechanisms in Si and pseudomorphic SiGe p-channel field-effect transistors, *Appl. Phys. Lett.* **85**, 6019-6021 (2004).
24. A. Kerber, E. Cartier, L. Pantisano, R. Degraeve, T. Kauerauf, Y. Kim, A. Hou, G. Groeseneken, H. E. Maes, and U. Schwalke, Origin of the threshold voltage instability in SiO_2/HfO_2 dual layer gate dielectrics, *IEEE Electron Device Lett.* **24**, 87-89 (2003).
25. C. Leroux, J. Mitard, G. Ghibaudo, X. Garros, G. Reimbold, B. Guillaumot, and F. Martin, Characterization and modelling of hysteresis phenomena in high k dielectrics, in *IEDM Tech. Dig.*, 2004, pp. 737-740.
26. S. Jakschik, A. Avellan, U. Schroeder, and J. W. Bartha, Influence of Al_2O_3 dielectrics on the trap-depth profiles in MOS devices investigated by the charge-pumping method, *IEEE Trans. Electron Devices* **51**, 2252-2255 (2004).

PROBLEMS

1. The circuit in Fig. 2-12 below is a setup to measure the noise in an unknown sample with resistance R which shows $1/f$ noise with a current noise PSD $S_{I,1/f}$ and (of course) thermal noise. The resistor R_V only shows

2. Noise characterization

thermal noise. Give an expression for the voltage noise at the output of the voltage amplifier (amplification A) including all noise sources. The input impedance of the amplifier is much larger than R or R_V. The internal noise of the amplifier must be taken into account.

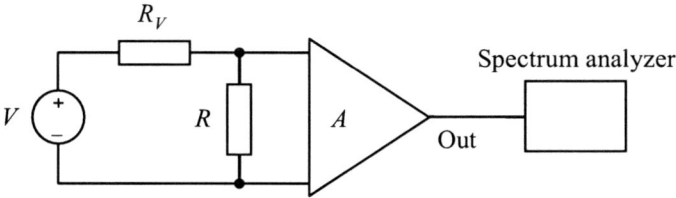

Figure 2-12. Measurement setup in problem 1.

2. Calculate the relative contribution from the $1/f$ noise in the S/D resistance to the output drain current noise for $I_D = I_{D,1}/3$. The noise contributions at the output from the channel and the S/D resistance are equally strong at $I_{D,1}$. Assume that the channel noise is of the number fluctuation type. The device is biased in the linear regime at a constant V_{DS}.

3. Derive Eq. (2-11)

$$N < 1/4\pi\alpha_H$$

where α_H is the Hooge parameter for the $1/f$ noise and N is the number of carriers (for example in the channel of a MOSFET).

4. Make an estimation under what conditions (gate area and $1/f$ noise level) RTS noise can be observed in a MOSFET with oxide trap density $N_t = 1\times10^{17}$ cm^{-3}eV^{-1} and 2 nm gate oxide thickness.

Chapter 3

1/f NOISE IN MOSFETS
Origins and modeling

1. INTRODUCTION

The CMOS technology has seen a rapid development in the past decades that made it possible to downscale the transistor dimensions at an exponential rate over time following the well-known Moore's law.[1] The transistor speed and the number of transistors that can be crammed into one chip have greatly increased as a result of the miniaturization of the transistor size. This evolution has stimulated the recent explosion in information and communication technology. CMOS technology has made inroads in the RF and analog domain that was previously dominated by bipolar transistors, which has created new demands on the MOS transistors.[2-7] Low 1/f noise in the transistors is an important requirement for low-noise RF/analog applications. Accurate MOSFET noise models are therefore of high relevance for CMOS circuit designers as well as semiconductor manufacturers and device designers need to be concerned about reducing the 1/f noise in the MOSFETs.

In order to be able to reduce the 1/f noise in MOS transistors and derive accurate noise models, one need to understand the 1/f noise mechanisms. This chapter is devoted to discussing the 1/f noise mechanisms in MOSFETs and the corresponding modeling of the noise sources. The origin of the 1/f noise in MOS transistors has been debated for several decades, whether number fluctuation noise due to traps in the gate oxide or bulk mobility fluctuations dominate the 1/f noise. The drain current in a MOSFET is confined to a narrow surface channel under the gate oxide. The current transport is therefore sensitive to traps present at the interface. Number fluctuations is generally believed to be the dominant $1/f$ noise mechanism in n-channel MOSFETs and commonly also in pMOSFETs.[8,9] However, the mobility fluctuation noise model tends to be better to explain the $1/f$ noise in pMOS transistors.[10-12] We will discuss these two mechanisms, number and mobility fluctuations, in section 3 and 4, respectively. We will begin by describing the equivalent noise circuit of a MOSFET and give a general overview of the noise sources in section 2. In section 5, the $1/f$ noise dependence on the voltage on the bulk terminal (substrate bias) is reported and discussed. This effect was recently discovered and is not implemented in the standard $1/f$ noise models.[13-14] The compact noise models are briefly reviewed in section 6, and the final section 7 deals with input referred gate voltage noise.

2. MOSFET NOISE MODEL

2.1 MOSFET fundamentals

2.1.1 Current-voltage relationships

We will first briefly review the MOSFET fundamentals before we describe the MOSFET noise sources and the equivalent noise circuit in the next subsection. A MOS transistor has four terminals, see Fig. 3-1. A voltage on the gate terminal (input) controls the current flowing between the source and drain (output) terminals. The substrate terminal is usually connected to ground, only with a small leakage current flowing through it. The source and drain regions are heavily doped and of opposite type than the substrate. For an n-channel MOSFET, which is exemplified in Fig. 3-1, source/drain is n^+-doped and the substrate p-type. The gate electrode, usually made of metal or poly-silicon, is separated from the Si substrate by a thin insulating film (thickness t_{ox}) called the gate oxide or gate dielectric. SiO_2 or nitrided SiO_2 is typically used as gate dielectrics in production today, but other materials with higher dielectric constant such as HfO_2 have been heavily researched and will likely replace SiO_2 in the future, as will be discussed in chapter 4. The intrinsic part of the MOSFET, the channel, is separated from the S/D terminals by a resistive extrinsic part with resistances R_S and R_D ($R_{SD} = R_S + R_D$). The S/D regions are heavily doped to make the S/D resistances as small as possible. However, the S/D resistances will limit the drain current for short gate lengths and is therefore one of the most difficult technological problems in realizing high-performance sub-100 nm gate length devices.

3. 1/f noise in MOSFETs

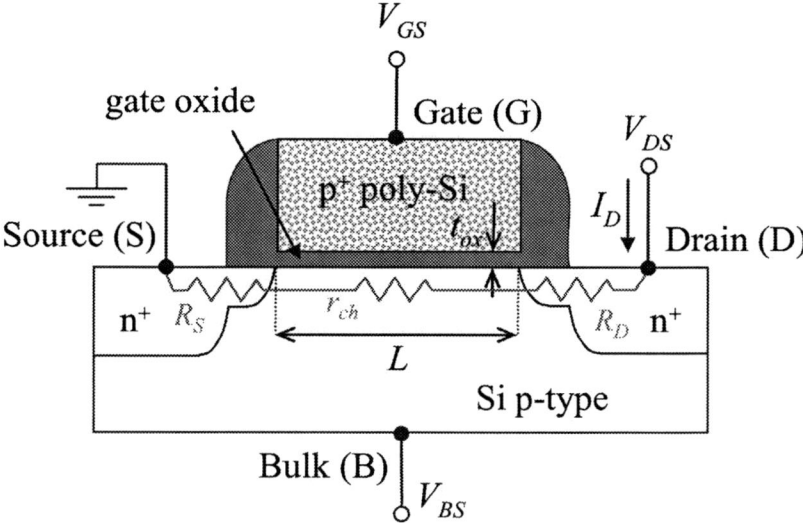

Figure 3-1. A schematic cross section of a MOSFET defining the terminal voltages and currents and some important transistor dimensions.

When the gate voltage is lower than a voltage level called the threshold voltage, only a small leakage current can flow between source and drain. The n^+-p-n^+ structure can be viewed as two p-n diodes connected back-to-back, preventing a current to flow except a small diffusion current. Biasing the gate with a positive voltage will increase the surface potential and repel holes from the surface leaving a negative charge of depleted ionized dopants. Increasing the gate voltage above the threshold voltage will *invert* the substrate and a channel of carriers of the opposite type (electrons in this case) is formed at the interface between the SiO_2 and Si substrate. The formation of the channel allows a large current to flow between source and drain, the device is switched on. In inversion, the inversion charge density can be approximated as

$$Q_i(V) = C_{ox}(V_{GS} - V_T - mV) \qquad (3\text{-}1)$$

where V is the potential along the channel, V_T is the threshold voltage, m is a body-effect coefficient, and $C_{ox} = \varepsilon_{ox}/t_{ox}$ is the oxide capacitance per unit area. The gate voltage controls the charge density in the channel, thereby modulating the conductivity between drain and source. The *drain current* between drain and source depends on the conductivity and the applied electric field along the channel. The drain current can be derived in the linear (triode) region where $V_{DS} < V_{DS,sat}$ as

$$I_D = \frac{W}{L}\mu_{eff}C_{ox}\left[(V_{GS} - V_T)V_{DS} - mV_{DS}^2/2\right]. \tag{3-2}$$

The threshold voltage is given by:

$$V_T = V_{fb} \pm 2\psi_B \pm \frac{\sqrt{4\varepsilon_{si}qN_{sub}\psi_B}}{C_{ox}} \tag{3-3}$$

where N_{sub} is the doping concentration in the substrate, V_{fb} the flat-band voltage, ε_{Si} the permittivity of Si and ψ_B the energy difference between the Fermi level E_F and the intrinsic level E_i. The plus signs in Eq. (3-3) apply for nMOS and the minus signs for pMOS, respectively.

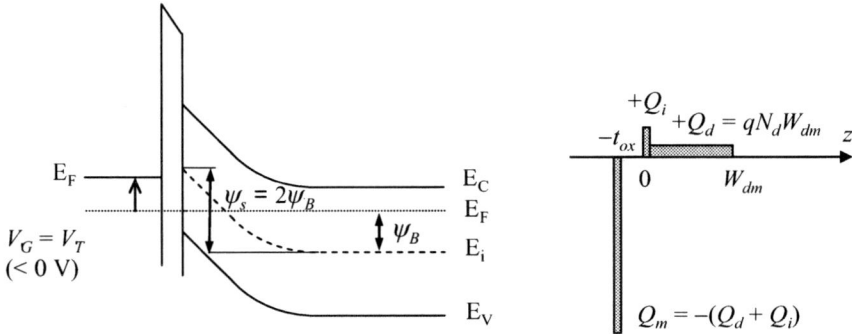

Figure 3-2. Energy band diagram illustration of a pMOSFET biased at threshold. The picture to the right is a schematic description of the charge distribution in the MOS structure.

The threshold voltage is the voltage required to achieve a surface potential (band bending) $\psi_s = 2\psi_B$ as described in Fig. 3-2 (shown for a pMOS). The flat-band voltage depends on the work function difference between the gate material and the substrate material ϕ_{ms} and the equivalent (trapped or fixed) oxide charge density at the oxide-silicon interface Q_{ox}

$$V_{fb} = \phi_{ms} - Q_{ox}/C_{ox}. \tag{3-4}$$

For n^+ or p^+ doped poly-Si gate and p-type or n-type Si substrate, the work function difference is calculated to be

$$\phi_{ms} = \pm 0.56 \pm kT/q \cdot \ln(N_{sub}/n_i) \tag{3-5}$$

where the plus or minus sign in front of the first term is for p-type or n-type gate material, respectively. For the second term, the plus sign applies for n-type substrate and the minus sign for p-type.

The factor m in Eqs. (3-1) and (3-2), called the body-effect coefficient, has been inserted to account for corrections to the simple theory. The value of m is typically between 1 and 1.4 and is calculated as follows:[15]

$$m = 1 + \frac{\sqrt{\varepsilon_{si} q N_{sub} / 4\psi_B}}{C_{ox}}. \tag{3-6}$$

The drain current in Eq. (3-2) increases with the drain voltage until a maximum is reached and saturation occurs. The drain voltage at saturation is

$$dI_D / dV_{DS} = 0 \Rightarrow V_{DS,sat} = (V_{GS} - V_T) / m. \tag{3-7}$$

At that point, called pinch-off, the channel at the drain end vanishes. The electric field along the channel between the source end and the pinch-off point stays constant with increasing $V_{DS} > V_{DS,sat}$ resulting in essentially the same current $I_{DS,sat}$. By inserting Eq. (3-7) in Eq. (3-2) the drain current in the saturation region can be written as

$$I_D = \frac{W}{L} \mu_{eff} C_{ox} \frac{(V_{GS} - V_T)^2}{2m}. \tag{3-8}$$

The pinch-off point moves slightly towards the source side for $V_{DS} > V_{DS,sat}$, which decreases the effective channel length somewhat. This effect is called *channel length modulation* and results in a weak increase of I_{DS} with V_{DS} in saturation.

The current will not go to zero when biased below threshold, $V_{GS} < V_T$, called the subthreshold region. A small diffusion current will remain

$$I_D = \mu_{eff} C_{ox} \frac{W}{L} (m-1) \left(\frac{kT}{q}\right)^2 e^{q(V_{GS}-V_T)/mkT} \left(1 - e^{-qV_{DS}/kT}\right). \tag{3-9}$$

The ability to turn off a device is described by the subthreshold slope

$$SS = \left(\frac{d(\log_{10} I_D)}{dV_{GS}}\right)^{-1} \approx 2.3 \frac{mkT}{q} = 2.3 \frac{kT}{q}\left(1 + \frac{C_{dm}}{C_{ox}}\right). \tag{3-10}$$

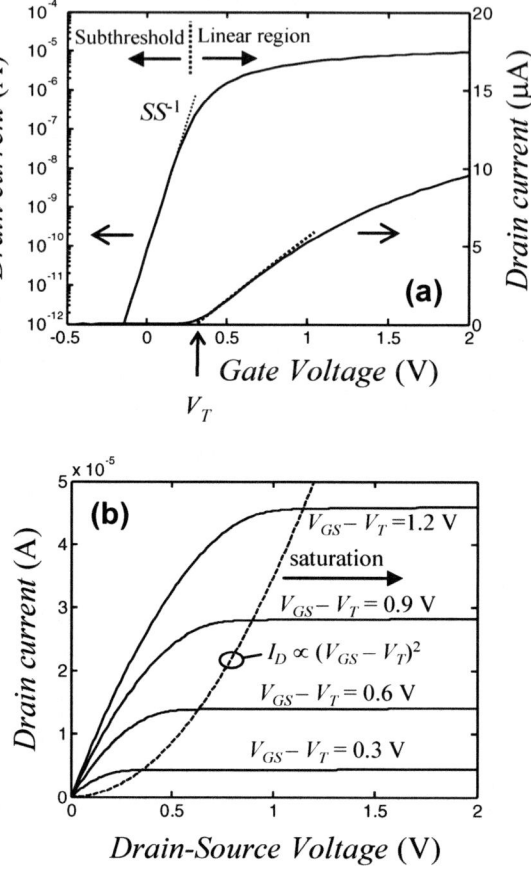

Figure 3-3. (a) I_D-V_{GS} characteristics in both logarithmic (left) and linear (right) scales. (b) I_D-V_{DS} characteristics for four different gate voltage overdrives.

A low subthreshold slope (*SS*) is desired since the current drops steeper with decreasing gate voltage, the device is easier to turn off. This allows a lower threshold voltage and consequently a higher on-current. An ideally low *SS* value of ~60 mV/dec can be achieved in SOI-technology whereas *SS* typically is between 60 and 100 mV/dec in bulk Si MOSFETs. The subthreshold slope is sensitive to the presence of traps at the SiO_2/Si interface since the capacitance associated with the interface states will act in parallel with the (maximum) depletion-layer capacitance C_{dm} and thus increase *SS*. Finally, the drain current characteristics are shown for the different regions of operation in Fig. 3-3, where the I_D-V_{GS} characteristics are displayed in (a) and the I_D-V_{DS} characteristics in (b).

2.1.2 Carrier mobility

The charge carriers in a semiconductor, which are placed under thermal equilibrium and with no electric field applied, move rapidly with the thermal velocity $\sim 10^7$ cm/s in random directions with no net current flow. The carriers are scattered by lattice vibrations (phonons) and impurities (dopants or defects) whereby their velocities are abruptly changed, under conservation of energy and momentum. The time between scattering events, the collision time τ_c, is typically on the order of 0.1 ps.[15] The carriers are accelerated between the collisions when they are under influence of an electric field. The carriers are assumed to immediately relax upon a collision and emerge at a random direction and with a speed corresponding to the local temperature. Therefore, at a certain point of time, the carriers will on average have been accelerated by the force qE during the time τ_c and gained a drift speed $v_d = \tau_c qE/m^*$, where m^* is the effective mass. Holes move in the same direction as the field and electrons in the opposite direction. One defines the mobility according to $\mu = v_d/E$, which thus equals $\mu = q\tau_c/m^*$. The carrier mobility in an inversion layer of a MOSFET is lower than in the bulk since the carriers are confined to a narrow region below the oxide/substrate interface and therefore suffer from scattering at the surface (roughness and surface phonons). By assuming that the different scattering mechanisms act independently and have the same energy dependence, the effective mobility μ_{eff} in an inversion layer of a MOSFET can be computed using Matthiessen's rule from the individual mobilities according to[16-19]

$$\frac{1}{\mu_{eff}} = \frac{1}{\mu_b} + \frac{1}{\mu_{ac}} + \frac{1}{\mu_{sr}} + \frac{1}{\mu_C} \qquad (3\text{-}11)$$

where μ_b is the bulk phonon mobility, μ_{ac} the mobility limited by surface acoustic phonon scattering, μ_{sr} the mobility due to surface roughness scattering and μ_C the mobility limited by Coulomb scattering mainly from ionized impurities and fixed/trapped charge in the gate oxide (its bulk and surfaces) or, if a very thin gate oxide is used, also from depleted charge in the poly-Si gate.[20] Although the conditions for using Matthiessen's rule seldom are fulfilled in practice, the formula still serves as a good approximation for the effective mobility.

The mobility in Si MOSFETs has been investigated extensively and the different scattering sources are well understood. The different scattering mechanisms depend in different ways on the effective electric field and the temperature. Fig. 3-4 shows a schematic diagram of the E_{eff} dependencies and describes how the different scattering mechanisms generally affect the mobility.[17]

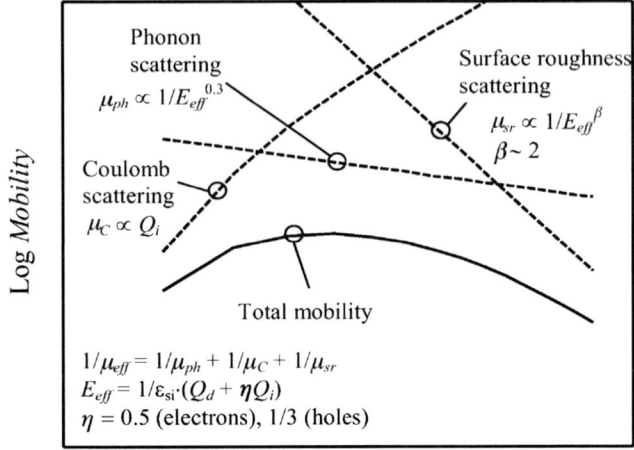

Figure 3-4. Schematic description of the E_{eff} and Q_i dependencies of the different scattering mechanisms and how they affect the mobility in an inversion layer of a MOSFET.

Phonon scattering and Coulomb scattering are both strongly temperature dependent, whereas surface roughness scattering shows a weaker dependence on temperature. Since the lattice vibrations increase with increasing temperature (more phonons are excited), the phonon limited mobility decreases. The thermal velocity of the carriers increases by increasing temperature. The carriers will then have a shorter interaction with the charged impurities resulting in reduced Coulomb scattering. For bulk semiconductors, temperature relations according to $\mu_b \propto T^{-3/2}$ and $\mu_C \propto T^{3/2}$ have been observed.[21]

2.2 MOSFET noise sources

2.2.1 Noise equivalent circuit

A MOSFET is a complex device containing purely resistive parts and a channel whose conductance is controlled by the gate voltage. Usually, the $1/f$ noise at the output is generated in the channel, but the $1/f$ noise originating from the S/D resistance contributes and may even take over as the dominant source at high drain currents. The low-frequency noise equivalent circuit of a MOSFET is shown in Fig. 3-5. For a short-circuited output, the total output drain current noise PSD from the uncorrelated noise sources in the channel and the S/D regions can be expressed as

$$S_{I_{D_{tot}}} = \frac{S_{I_{D,ch}} + g_{ch}^2 R_D^2 S_{I_{RD}} + R_S^2(g_m + g_{ch})^2 S_{I_{RS}}}{[1 + g_m R_S + g_{ch}(R_S + R_D)]^2}. \tag{3-12}$$

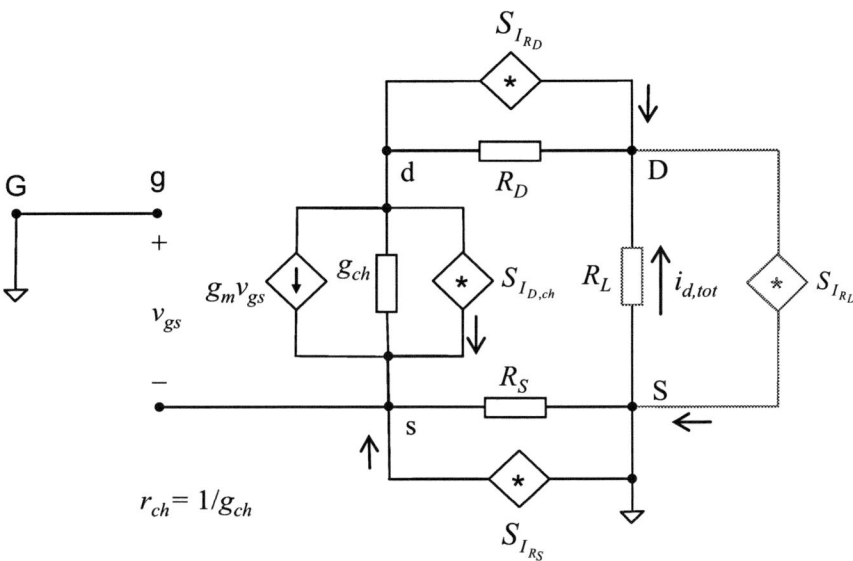

Figure 3-5. Small-signal equivalent circuit of a MOSFET including noise sources.

For the general case with a load resistance R_L at the output, the output drain current noise is

$$S_{I_{D_{tot}}} = \frac{S_{I_{D,ch}} + g_{ch}^2 R_D^2 S_{I_{RD}} + R_S^2(g_m + g_{ch})^2 S_{I_{RS}}}{[1 + g_m R_S + g_{ch}(R_S + R_D + R_L)]^2} + \frac{S_{I_{RL}}(1 + g_m R_S + g_{ch}(R_S + R_D))^2}{[1 + g_m R_S + g_{ch}(R_S + R_D + R_L)]^2}. \tag{3-13}$$

It is instructive to derive the expression for the total drain current noise in Eq. (3-13). Since the equivalent circuit contains a dependent current generator, the superposition principle is not advisable to use. Instead, we start the derivation by writing the noise current $i_{d,tot}$ through R_L

$$i_{d,tot} = i_{d,ch} + g_m v_{gs} + g_{ch} v_{ds} + i_{R_L}. \tag{3-14}$$

Here, g_m is the transconductance, defined as

$$g_m = \partial I_D / \partial V_{GS}. \tag{3-15}$$

The gate-source voltage fluctuates due to the noise voltage over R_S

$$v_{gs} = -R_S(i_{d,tot} + i_{R_S} - i_{R_L}). \tag{3-16}$$

The drain-source voltage fluctuations are found by applying Kirchoff's voltage law around the loop

$$v_{ds} = -R_S(i_{d,tot} + i_{R_S} - i_{R_L}) - R_L i_{d,tot} - R_D(i_{d,tot} + i_{R_D} - i_{R_L}). \tag{3-17}$$

By solving Eqs. (3-14), (3-16) and (3-17) the noise current $i_{d,tot}$ can be written as

$$i_{d,tot} = \frac{i_{d,ch} - i_{R_S}(g_m R_S + g_{ch} R_S) - i_{R_D} g_{ch} R_D}{1 + g_m R_S + g_{ch}(R_S + R_D + R_L)} + \frac{i_{R_L}(1 + g_m R_S + g_{ch} R_S + g_{ch} R_D)}{1 + g_m R_S + g_{ch}(R_S + R_D + R_L)}. \tag{3-18}$$

Note that all noise current contributions from each source are grouped together. Since all noise sources are uncorrelated, the PSD in Eq. (3-13) is found by squaring each term in Eq. (3-18) and replacing the noise currents with their corresponding PSDs.

The drain current noise is a superposition of several noise sources with different spatial location and with different physical origins. The lower limit of the noise is always (white) thermal noise or shot noise. On top of the white noise, $1/f$ noise is usually present, and g-r noise can sometimes also be observed especially in MOSFETs with a very small gate area (~0.1 μm^2). The channel $1/f$ noise due to the number and mobility fluctuation noise mechanisms are discussed in sections 3 and 4 of this chapter, respectively. In the next subsection, we discuss the modeling of the noise in the S/D resistances and the channel thermal noise.

2.2.2 Channel thermal noise and noise from extrinsic regions

The noise originating from the source and drain resistances can be modeled as a sum of thermal noise and mobility fluctuation $1/f$ noise

3. 1/f noise in MOSFETs

$$S_{I_{RS}} = \frac{\alpha_{H,S} I_D^2}{fN} + 4kT/R_S .\qquad(3\text{-}19)$$

Here, N is the number of carriers in the source region. Of course, the noise from the drain resistance is modeled in the same way. Usually, the source and drain regions are symmetrical which means that the $R_S = R_D$ and $\alpha_{H,S} = \alpha_{H,D}$.

The channel is resistive in the linear and saturation regions and therefore generates thermal noise. In subthreshold regime, on the other hand, shot noise with a drain current noise PSD equal to $2qI_D$ is generated. The thermal noise in the channel depends on the operating condition according to[22]

$$S_{I_{D,chth}} = 4kT\left[\frac{W}{L}\mu_{eff} C_{ox}(V_{GS} - V_T)\frac{2}{3}\frac{1+\eta_v+\eta_v^2}{1+\eta_v}\right]\qquad(3\text{-}20)$$

where η_v is defined as

$$\eta_v = \begin{cases} 1 - V_{DS}/V_{DS,sat}, & V_{DS} \le V_{DS,sat} \\ 0, & V_{DS} > V_{DS,sat} \end{cases} \quad \text{where } V_{DS,sat} = \frac{(V_{GS}-V_T)}{m}.\quad(3\text{-}21)$$

Thus, in the linear region at a small V_{DS}, $\eta_v \approx 1$

$$S_{I_{D,chth}} = 4kT\left[\frac{W}{L}\mu_{eff} C_{ox}(V_{GS} - V_T)\right] = 4kTg_{ch}\qquad(3\text{-}22)$$

and in the saturation region where $\eta_v = 0$

$$S_{I_{D,chth}} = 4kT\left[\frac{W}{L}\mu_{eff} C_{ox}(V_{GS} - V_T)\frac{2}{3}\right] = 4kTg_{ch,0}\gamma\qquad(3\text{-}23)$$

($= 4kTg_m\gamma$ for $m = 1$).

Here $g_{ch,0}$ is the channel conductance at zero drain-source bias. From theory, the coefficient γ equals 2/3. However, a more detailed derivation shows that γ can be slightly larger than 2/3 for short channels (below ~0.5 μm) due to velocity saturation and channel length modulation.[23] The coefficient γ can also show a slight increase for long channel devices (above ~1 μm) as well due to the nonquasi-static effect. More importantly, when the drain voltage is high enough to cause avalanche multiplication of carriers, the thermal noise can increase significantly. It is found that[23]

$$\gamma = \frac{2}{3}M^2 - \frac{4}{3}(M-1) + \frac{2qI_S M(M-1)}{4kTg_{ch,0}} \qquad (3\text{-}24)$$

in case of avalanche where $M (= I_D/I_S)$ is the multiplication factor. Note that in some work,[24,25] very large γ values above 10 have been found in experiments which have been claimed to be due to hot carriers (local temperature higher than 300 K). From theoretical modeling, however, hot carriers are not believed to cause a significant increase of the noise.[23,26]

3. NUMBER FLUCTUATIONS

In this section, we will discuss $1/f$ noise generated in the channel of the MOSFET due to the number fluctuation noise mechanism. The drain current in a MOSFET is, as mentioned previously in this chapter, confined to a narrow region under the gate oxide surface, called the channel. While a superposition of g-r noise spectra to produce $1/f$ noise is unlikely to occur for a homogenous bulk device since the required distribution of time constants is not possible to achieve without very special assumptions, this can easily be obtained for a surface channel. In 1957, McWorther presented a $1/f$ noise model based on quantum mechanical tunneling transitions of electrons between the channel and traps in the gate oxide.[27] The tunneling time varies exponentially with distance, thus the required distribution of time constants to produce $1/f$ noise is obtained for a trap density that is uniform in both energy and distance from the channel interface. The McWorther model is celebrated for its simplicity and excellent agreement with experiments, especially for nMOS transistors.[8,9] The $1/f$ noise in pMOS transistors, on the other hand, is often better explained by the mobility fluctuation noise model.[10-12,28] Also note that there are a few reports about mobility fluctuation noise in nMOSFETs.[29,30] It was later observed that a trapped carrier also affects the surface mobility through Coulomb interaction. The so-called correlated mobility fluctuations gave a correction to the number fluctuation noise model that was suggested to resolve the deviations found in pMOSFETs.[31] However, the correction factor was criticized for being unphysically high since screening was not accounted for.[32]

We will in the following subsections analyze the number fluctuations and the McWorther model as well as discuss the impact of correlated mobility fluctuations.

3.1 Number fluctuation noise

The physical mechanism behind the number fluctuation $1/f$ noise in MOSFETs is the interaction between slow traps in the gate oxide and the carriers in the channel, which is schematically illustrated in Fig. 3-6. The oxide traps dynamically exchange carriers with the channel causing a fluctuation in the surface potential, giving rise to fluctuations in the inversion charge density. If a drain current is flowing in the device, these fluctuations are translated to the current which can be observed from measurements. The fluctuating oxide charge density δQ_{ox} is equivalent to a variation in the flat-band voltage, see Eq. (3-4).

$$\delta V_{fb} = -\delta Q_{ox} / C_{ox}. \tag{3-25}$$

The fluctuation in the drain current $I_D = f(V_{fb}, \mu_{eff})$ then yields[33]

$$\delta I_D = \frac{\partial I_D}{\partial V_{fb}} \delta V_{fb} + \frac{\partial I_D}{\partial \mu_{eff}} \frac{\partial \mu_{eff}}{\partial Q_{ox}} \delta Q_{ox}. \tag{3-26}$$

Since $\partial I_D / \partial V_{fb} = -\partial I_D / \partial V_{GS} = -g_m$ ($+g_m$ for pMOS) and $I_D \propto \mu_{eff}$

$$\delta I_D = -g_m \delta V_{fb} + \frac{I_D}{\mu_{eff}} \frac{\partial \mu_{eff}}{\partial Q_{ox}} \delta Q_{ox}. \tag{3-27}$$

One can define a coupling parameter or scattering parameter that reflects how a variation in the oxide charge couples to the mobility

$$\alpha = \frac{1}{\mu_{eff}^2} \frac{\partial \mu_{eff}}{\partial Q_{ox}} \quad \text{(valid for nMOS, minus sign for pMOS)}. \tag{3-28}$$

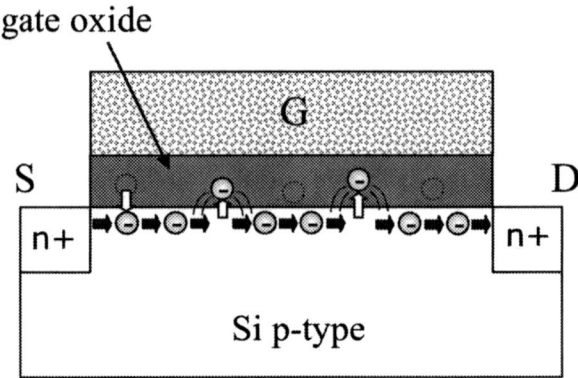

Figure 3-6. Schematic illustration of electrons in the channel of a MOSFET moving in and out of traps, giving rise to fluctuations in the inversion charge density and thereby the drain current. The carrier mobility is also affected by the oxide charge; mobility fluctuations *correlated* to the number fluctuations are therefore generated which may increase or decrease the total LF noise.

Inserted in Eq. (3-27) this gives

$$\delta I_D = -g_m \delta V_{fb} - I_D \mu_{eff} \alpha C_{ox} \delta V_{fb}. \tag{3-29}$$

Calculating the power spectral density

$$S_{I_D} = S_{V_{fb}} \left(1 + \frac{\alpha \mu_{eff} C_{ox} I_D}{g_m} \right)^2 g_m^2. \tag{3-30}$$

The first term in the parentheses in Eq. (3-30) is due to fluctuating number of inversion carriers and the second term to mobility fluctuations correlated to the number fluctuations. Note that α can be negative or positive depending on if the mobility increases or decreases upon trapping a charge according to Eq. (3-28).

3.2 The McWorther model

Eq. (3-30) is a general expression that was derived without any assumptions about the exact mechanism behind the fluctuations in the flat-band voltage. Now, we will derive an expression under the condition that tunneling transitions of carriers into and out of traps in the gate oxide is the origin of the number fluctuations. This is in essence the theory of A. L.

3. 1/f noise in MOSFETs

McWorther,[27] which has a widespread recognition as the principal explanation of 1/f noise in MOSFETs.

The g-r noise in the oxide charge generated by one trap that randomly captures and releases channel electrons can be written as

$$S_{Q_{ox}} = S_{V_{fb}} C_{ox}^2 = \frac{q^2}{W^2 L^2} 4\overline{\Delta N_{OX}^2} \frac{\tau}{1+(2\pi f \tau)^2}. \tag{3-31}$$

The variance in ΔN_{OX}, where N_{OX} is the number of oxide charges, due to a trap at energy E is calculated from the probability that the trap is occupied. The probability is given by the Fermi-Dirac distribution function $f(E)$

$$f(E) = \left[1 + e^{(E-E_{F,n(p)})/kT}\right]^{-1}. \tag{3-32}$$

Then

$$\overline{\Delta N_{OX}^2} = (1-f(E))^2 f(E) + (0-f(E))^2 (1-f(E)) = f(E)(1-f(E)). \tag{3-33}$$

Now, the contributions from all traps in the gate oxide should be taken into account. The total PSD is found by summing over all traps whose noise contributions are given by Eq. (3-31). However, the individual traps are not known. We instead assume a density of traps $N_t(x,y,z,E)$ in volume and energy and make an integration:[31,34]

$$S_{Q_{ox}} = \frac{q^2}{W^2 L^2} \int_{E_V}^{E_C} \int_0^W \int_0^L \int_0^{t_{ox}} 4 N_t f(E)(1-f(E)) \frac{\tau}{1+(2\pi f \tau)^2} dx\,dy\,dz\,dE. \tag{3-34}$$

The product $f(E)(1-f(E)) = -kT\, df(E)/dE$ is sharply peaked around the quasi-Fermi level and N_t is considered as uniform over the gate area. Thus

$$S_{Q_{ox}} = \frac{q^2 kT}{WL} \int_0^{t_{ox}} 4 N_t \frac{\tau}{1+(2\pi f \tau)^2} dz \tag{3-35}$$

where N_t is the density of traps in the gate dielectrics at the quasi-Fermi level (in $cm^{-3} eV^{-1}$) since these traps are the only ones that contribute to the 1/f noise. Other traps are permanently filled or permanently empty. In the

McWorther model, which assumes that trapping and detrapping occur through tunneling processes, the trapping time constant is given as

$$\tau = \tau_0(E) \cdot e^{z/\lambda} \tag{3-36}$$

for an electron tunneling from the interface ($z = 0$) to a trap located at a distance z in the gate oxide. The tunneling attenuation length λ is predicted by the Wentzel-Kramers-Brillouin (WKB) theory to be[34]

$$\lambda = \left[\frac{4\pi}{h} \sqrt{2m^* \Phi_B} \right]^{-1} \tag{3-37}$$

where Φ_B is the tunneling barrier height seen by the carriers at the interface and h is Planck's constant. Calculations using Eq. (3-37) give $\lambda \approx 1$ Å for the Si/SiO$_2$ system. The time constant τ_0 is often taken as 10^{-10} s. This yields $z = 2.6$ nm and 0.7 nm for a frequency of 0.01 Hz and 1 MHz, respectively. Thus, oxide traps located too close to the channel interface are too fast to give $1/f$ noise, and those located more than ~3 nm from the interface are too slow to contribute. By inserting Eq. (3-36), the integral in Eq. (3-35) can be evaluated as (see chapter 1.3.5)

$$S_{V_{fb}} = \frac{q^2 kT\lambda N_t}{f^\gamma WLC_{ox}^2}. \tag{3-38}$$

The frequency exponent γ deviates from 1 if the trap density is not uniform in depth; $\gamma < 1$ is expected when the trap density is higher close to the gate oxide/channel interface than that in the interior of the gate oxide, and $\gamma > 1$ for the opposite case.[35] One example, which is a good evidence of the McWorther model, is the observation of $1/f^{1.7}$ noise.[36] The devices in that particular experiment had a nitrided gate oxide, which contains a higher density of traps than pure thermally grown SiO$_2$. If the peak of the nitrogen profile is located away from the channel interface (due to re-oxidation for example), the observation of $\gamma = 1.7$ can be explained.

The bias dependence of the normalized drain current noise in the number fluctuation model is simulated for drain currents ranging from subthreshold to strong inversion regimes (at small V_{DS}) using Eq. (3-30) with $\alpha = 0$, 1.5×10^4 Vs/C and -1.5×10^4 Vs/C and a constant arbitrary N_t (other parameters from a standard Si pMOSFET), the result is shown in Fig. 3-7. Normalization by dividing the current noise PSD with the current squared is often performed in noise analysis since S_I/I^2 is inversely proportional to the signal-to-noise ratio.

3. 1/f noise in MOSFETs

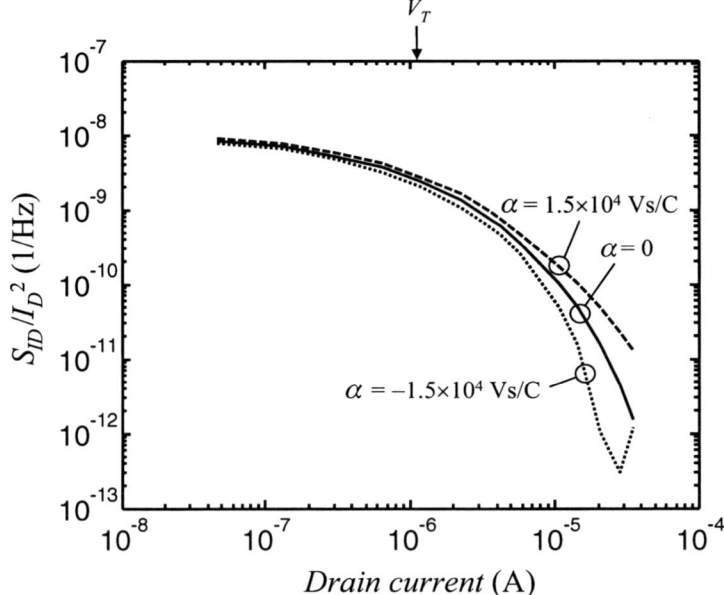

Figure 3-7. Simulation of the normalized drain current at small V_{DS} using the number fluctuation noise model in Eq. (3-30) with $\alpha = 0$, 1.5×10^4 Vs/C and -1.5×10^4 Vs/C and a constant *arbitrary* N_t. The other parameters were taken from a standard Si pMOSFET.

The normalized drain current noise varies approximately as

$$S_{I_D} = \frac{q^2 kT\lambda N_t}{f^\gamma WLC_{ox}^2} \frac{I_D^2}{(V_{GS} - V_T)^2} \tag{3-39}$$

since $g_m = I_D/(V_{GS} - V_T)$ in the first instance (in the linear region). Eq. (3-39) overestimates the number fluctuation noise at high drain currents due to the fact that g_m is decreasing with gate voltage beyond $g_{m,max}$. In subthreshold, on the other hand, the normalized drain current noise is almost constant since $g_m = I_D q/mkT$ according to Eq. (3-9). The physical explanation is that $|\delta Q_i/\delta Q_{ox}| < 1$, the charge trapped in the oxide is not only supplied from the inversion charge but also from the depletion and interface trap charges. The normalized drain current noise can written as[37]

$$\frac{S_{I_D}}{I_D^2} = \frac{q^4 \lambda N_t}{kTf^\gamma WL(C_{ox} + C_d + C_{it})^2} \tag{3-40}$$

in the subthreshold region. Here, C_d is the depletion capacitance and C_{it} is the capacitance due to interface states.

The trap density can also vary with energy, which affects the bias and frequency dependence of the noise. The band diagram in Fig. 3-8 describes the tunneling transitions, (i) directly[34] or (ii) using interface traps as stepping stones,[38] from the Si to the gate oxide, and the window (z,E) of traps seen at a particular bias point (shaded area). The interface trap density often shows a U-shaped curve as function of energy in the bandgap with increasing values towards the conduction or valence band edges. If the oxide trap density follows the same behaviour, N_t is predicted to increase with gate bias since the quasi-Fermi level approaches E_C or E_V. Due to the band bending of the gate oxide, an energy dependent trap density should be accompanied by a frequency exponent $\gamma \neq 1$. Traps in the oxide interior are swept "faster" in energy than the traps at the channel interface. Thus, it is expected that $\gamma > 1$ and increasing with gate bias in the case of a trap density that increases towards the band edges.

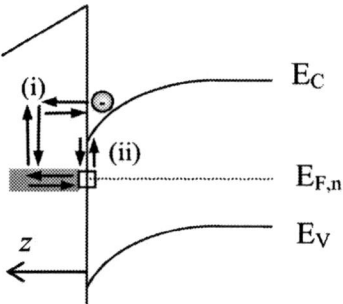

Figure 3-8. Energy band diagram showing the tunneling transitions of electrons between the conduction band and traps in the gate oxide, (i) corresponds to direct tunneling and (ii) to indirect tunneling via interface traps.

Finally, the tunneling model presented here is not the only one that can give the appropriate distribution of time constants in order to achieve the $1/f$ fluctuations. Another possibility is thermally activated traps with time constants exponentially depending on energy[39]

$$\tau_{th} = \tau_{0,th} e^{E/kT}. \tag{3-41}$$

$1/f$ noise is obtained for an even distribution of traps in energy. Studies on RTS noise in MOSFETs show that thermally activated phonon-assisted capture and emission of carriers play an important role.[40,41] Nevertheless, the support for the tunneling model is solid from experimental findings[8,36,42] but

3. 1/f noise in MOSFETs

thermally activated traps should also be considered in order to get a complete description of the number fluctuation noise.

3.3 Correlated mobility fluctuations

A trapped carrier will not only shift the flat-band voltage and thereby cause a fluctuation in the inversion charge density, the trapped carrier will also affect the mobility. These fluctuations in the mobility are correlated to the fluctuating inversion charge density, both related to the trapping and release of carriers in slow oxide traps. However, there is a disagreement in the literature regarding the strength of the correlated mobility fluctuations, which is usually modeled with a parameter α. A constant α is frequently used in the noise models, but this is not physically correct for several reasons such as the effect of screening.[32,43,44] Instead, α is expected to decrease with increasing inversion charge density Q_i due to the screening effect. The parameter α can be estimated both from low-frequency noise and mobility characterizations. For nMOSFETs, the α values obtained using these methods are usually consistent around 1×10^4 Vs/C. For pMOSFETs, on the other hand, much larger α values are often reported from noise characterizations, in the range 3-20 $\times 10^4$ Vs/C,[9,45-49] than expected from mobility models for pMOSFETs ($\alpha \sim$ 0.3-4 $\times 10^4$ Vs/C).[19,50] This large discrepancy suggests that the correlated mobility fluctuations in many cases are mistaken for another noise mechanism in pMOSFETs.

The disagreement in α values for pMOSFETs calls for a more detailed study. We will derive a model for α based on existing mobility models and compare with experiments. We start our analysis with the expression for α given in Eq. (3-28) and the mobility expression in Eq. (3-11). We perform the derivation for pMOSFETs here, but the same final expression is obtained for nMOSFETs. The Coulomb scattering is separated in a part caused by impurities and one part from charge in the oxide, $1/\mu_C = 1/\mu_{C,imp} + 1/\mu_{C,ox}$.

We can write

$$\alpha = -\frac{1}{\mu_{eff}^2}\frac{\partial \mu_{eff}}{\partial Q_{ox}} =$$
$$-\left(\frac{1}{\mu_{ac}^2}\frac{\partial \mu_{ac}}{\partial Q_i} + \frac{1}{\mu_b^2}\frac{\partial \mu_b}{\partial Q_i} + \frac{1}{\mu_{sr}^2}\frac{\partial \mu_{sr}}{\partial Q_i} + \frac{1}{\mu_{C,imp}^2}\frac{\partial \mu_{C,imp}}{\partial Q_i}\right)\frac{\partial Q_i}{\partial Q_{ox}} - \quad (3\text{-}42)$$
$$\frac{1}{\mu_{C,ox}^2}\frac{\partial \mu_{C,ox}}{\partial Q_i}\frac{\partial Q_i}{\partial Q_{ox}}.$$

The bulk phonon mobility μ_b is a constant and $\partial Q_i / \partial Q_{ox} = -1$ in strong inversion, which gives

$$\alpha = \frac{1}{\mu_{ac}^2} \frac{\partial \mu_{ac}}{\partial Q_i} + \frac{1}{\mu_{sr}^2} \frac{\partial \mu_{sr}}{\partial Q_i} + \frac{1}{\mu_{C,imp}^2} \frac{\partial \mu_{C,imp}}{\partial Q_i} + \frac{1}{\mu_{C,ox}^2} \frac{\partial \mu_{C,ox}}{\partial Q_i}. \qquad (3\text{-}43)$$

The first two terms will be negative since μ_{ac} and μ_{sr} fall off with increasing Q_i, whereas the third term will be positive. These terms are often neglected in the derivation of α. The last term will be positive or negative depending on the type of trap, channel type and the nature of the oxide charge; Table 3-1 summarizes the outcome for the different combinations.

Table 3-1. Sign of last term in Eq. (3-43).

	Acceptor trap (–/0)	Donor trap (0/+)
pMOS	–	+
nMOS	+	–

A simple model for $\mu_{C,ox}$ is[51]

$$\mu_{C,ox} = \frac{1}{\alpha_C |Q_{ox}|} \qquad (3\text{-}44)$$

where α_C is a Coulomb scattering parameter. A common approximation is to set $\alpha = \alpha_C$, which leads to an overestimation of α at high gate voltage overdrives as shown below. The magnitude of α_C decreases with increasing Q_i since the inversion charge screens the Coulomb interaction between the oxide charge and the carriers in the inversion layer. According to Vandamme and Vandamme

$$\alpha_C = \frac{1}{q \cdot \mu_{C0} \sqrt{C_{ox}(V_{GS} - V_T)/q}} \qquad (3\text{-}45)$$

where μ_{C0} is a fitting parameter given as 5.9×10^8 cm/Vs.[32] Another model for α_C that has been used in the literature is

$$\alpha_C = \alpha_0 - \alpha_1 \ln(N_s) \qquad (3\text{-}46)$$

where α_0 and α_1 are constants and $N_s = Q_i/q$ is the carrier density in the channel. According to the model by Pacelli *et al*, $\alpha_0 = 5.93 \times 10^5$ Vs/C and $\alpha_1 = 2.00 \times 10^4$ Vs/C for a trap located 0.5 nm inside the gate oxide.[52] If the trap instead is located 1.5 nm from the oxide/semiconductor interface, the

3. 1/f noise in MOSFETs

parameters values are reduced by almost 50%: $\alpha_0 = 3.12\times10^5$ Vs/C and $\alpha_1 = 1.06\times10^4$ Vs/C. The two aforementioned models were both developed for electrons. Unfortunately, corresponding data for holes are scarce in the literature. However, if we look for simpler models where α_C is treated as a constant some conclusions can be made. Dimitrijev and Stojadinovic found that $\alpha_C = 2.0\times10^4$ and 4.3×10^4 Vs/C for electrons and holes, respectively.[53] The value for holes is scaled by μ_{0n}/μ_{0p} in comparison with the value for electrons, where $\mu_{0n(p)}$ is the mobility of a nMOS (pMOS) transistor with no oxide charge. On the other hand, Emrani et al. report $\alpha_C = 3.4\times10^3$ Vs/C for holes, 3 times lower than the value for electrons (1.2×10^4 Vs/C) in their study.[50] As indicated above, the distance of the trap from the oxide interface determines the scattering strength, which could explain the disagreement in α values in the literature. We therefore assume that the models in Eqs. (3-45) and (3-46) can be used for holes as well, maybe with a scaling of the parameter values according to the ratio between the electron and hole mobility values.

The parameter α has been simulated using the Si inversion layer mobility model by Darwish and co-workers[54] and using the two aforementioned models for α_C in Eqs. (3-45) and (3-46) to determine the last term in Eq. (3-43). The result is plotted for electrons (an acceptor trap was assumed) in Fig. 3-9(a). Fig. 3-9(b) shows the result for holes, simulated using Eq. (3-45) with $\mu_{C0} = 5.9\times10^8$ cm/Vs and 1.5×10^8 cm/Vs (scaled with 1/4) under the assumption of a donor trap. The first three terms in Eq. (3-43) account for the difference between the solid lines and the broken lines in Fig. 3-9, which is around 4×10^3 Vs/C for holes and 2×10^3 Vs/C for electrons at high inversion carrier densities. The traps responsible for the 1/f noise are believed to be located 1-3 nm from the oxide/channel interface where the Coulomb interaction between the traps and the channel carriers is weaker than for interface traps. Therefore, one should rather use α values at the lower end of the observed range. A safe choice would be to assume that $|\alpha|$ is below $\sim 1\times10^4$ Vs/C, which means that the correlated mobility fluctuations make a quite small correction to the number fluctuation noise.

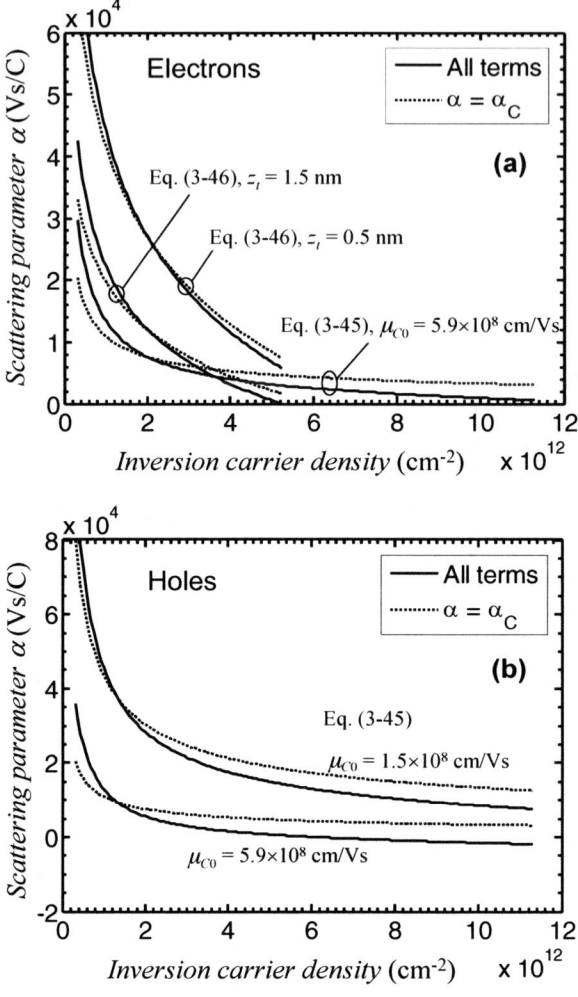

Figure 3-9. Scattering parameter α simulated for (a) electrons and (b) holes in a Si MOSFET using Eq. (3-43). The Coulomb scattering parameter α_C, the last term in Eq. (3-43), was modeled using Eq. (3-45) or (3-46). Only Eq. (3-45) was used for α_C in (b) but with two different values for the parameter μ_{C0}.

An experimental study aiming to characterize the parameter α from both mobility and low-frequency noise measurements on the same devices has been performed by the authors.[55] The devices used in the experiments were $Si_{0.8}Ge_{0.2}$ pMOSFETs with Al_2O_3 as gate dielectrics. The devices were subjected to water vapour annealing which modified the oxide charge in the devices. The response in mobility was measured as a function of the change in the oxide charge and was used to calculate α. The low-frequency noise in

3. 1/f noise in MOSFETs

the devices was measured as a function of drain current giving another independent method to measure α. The parameter α was in the latter case extracted from the LF noise data by a curve fitting to Eq. (3-30).

Fig. 3-10 shows the extracted α values from (a) the mobility measurements and (b) the LF noise measurements. As seen, the α values are in the same range for the two methods. Note that negative α values were found for un-annealed devices, meaning that the mobility increased when a carrier was trapped (acceptor trap for pMOS). After 210 min H_2O annealing α was instead found to be positive due to the neutralization of negative

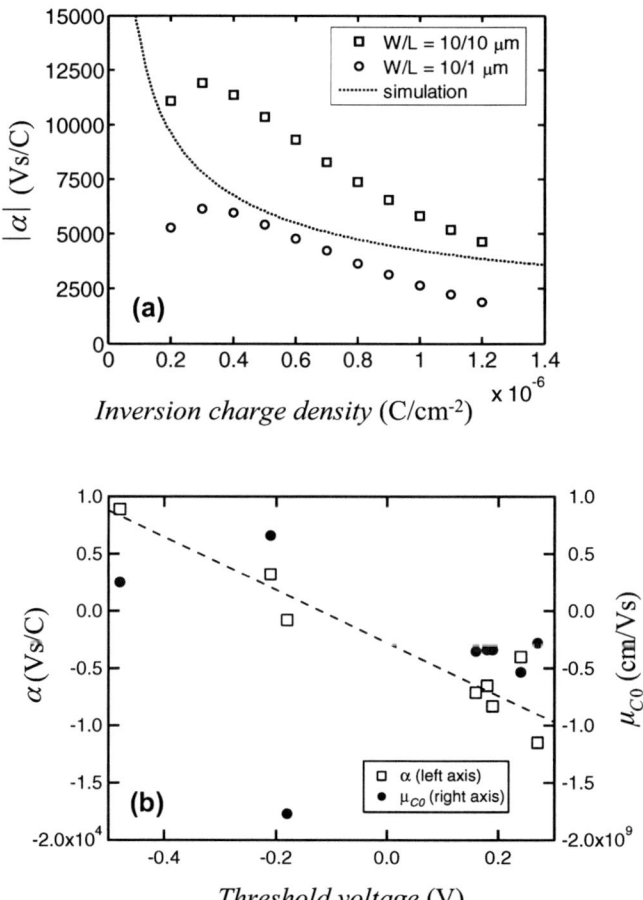

Figure 3-10. The scattering parameter α extracted from (a) mobility measurements and (b) LF noise measurements on $Si_{0.8}Ge_{0.2}$ pMOSFETs with Al_2O_3 gate dielectrics. The dotted line in (a) is a simulation using Eq. (3-45). Reprinted from von Haartman et al.[55] with permission from Elsevier.

charge or addition of positive charge induced by the annealing. The following model was suggested for α in case there are two types of traps[55]

$$\alpha_C = \frac{\alpha_{PC} N_{PC} - \alpha_{NC} N_{NC}}{N_{PC} + N_{NC}}. \qquad (3\text{-}47)$$

The subscripts *NC* and *PC* refer to traps giving a negative and positive correlation between ΔN and $\Delta \mu$, respectively. Finally, note that a Si MOSFET with SiO$_2$ as gate dielectrics can be expected to have roughly two times higher α values than those reported in Fig. 3-10 since

$$\alpha_C \propto \frac{m^*}{(\varepsilon_{sc} + \varepsilon_{ox})^2} \qquad (3\text{-}48)$$

according to theory.[31,56] Here, ε_{sc} and ε_{ox} are the permittivity values for the semiconductor and gate oxide, respectively, and m^* is the effective carrier mass in the semiconductor.

3.4   Critical discussion

The McWorther model often shows an excellent agreement with experimental 1/f noise data for nMOSFETs over the whole bias range from subthreshold to strong inversion. However, deviations from the expected behaviour are observed for pMOSFETs. In the framework of the McWorther model, two different theories have been suggested and investigated to explain the 1/f noise in the pMOSFETs. One such theory is the correlated mobility fluctuations, which was implemented in the model presented in Eq. (3-30). Values of the parameter α in the range 1-2 ×10^5 Vs/C are often necessary to model the 1/f noise in pMOSFETs.[9,45,46] However, these values are roughly an order of magnitude too high to be supported by the laws of physics as discussed in the previous subsection. By including screening and correct physical values,[32] the model in Eq. (3-30) often fails to describe the observed 1/f noise in pMOSFETs, except if N_t is allowed to vary with energy. A possible, but controversial, idea how to explain the bias dependence for pMOSFETs is to make the assumption that $N_t \propto V_{GT}$. Scofield et al. studied the noise versus temperature (77-300 K) and gate voltage in nMOS and pMOS transistors and claim that the number fluctuation noise model can be used for both device types.[57] The reason, according to the authors, is that the trap density is constant near the conduction band edge (nMOS), while it increases with energy close to the valence band edge. So far, this hypothesis has not been verified from experiments probing the same traps but using another independent technique. The so-called U-shaped distribution, where the trap density increases close to both the valence and conduction band edges, is typically obtained from different types of measurements of traps at the Si/SiO_2 interface.[58-60] However, the oxide traps and interface traps do not necessarily have the same origin. A consequence of an energy dependent N_t in the McWorther model is that the frequency exponent should vary with gate bias too due to the band bending in the oxide. This is, however, seldom analyzed or reported in relation with $N_t(E)$. Investigations of the gate voltage noise dependence on the oxide thickness t_{ox} in pMOSFETs by Knitel et al. showed that the gate voltage noise was proportional to $(t_{ox})^p$ with p ranging between 1.4 and 1.8 close to threshold and p around 1 at high V_{GT}.[61] This strongly points to a mobility fluctuation origin (Hooge noise or correlated mobility fluctuations) of the $1/f$ noise at high V_{GT}.

A type of noise experiment in favour of a number fluctuation origin is the occurrence of g-r noise in devices with small gate areas which adds up to $1/f$ noise in a large device. This has been shown by both Brederlow et al.[62] and Scholten et al.[23] It is, however, important to separate the $1/f$ noise and the g-r noise components, and make this study over a wide bias range in order to draw reliable conclusions. Another experiment pointing towards a trap origin of the $1/f$ noise is the switched bias experiments. When a rectangular pulse train is applied on the gate, a 5-8 dB reduction in the noise has been observed.[63] This can be explained by the number fluctuation noise model only if the trap densities differ at the two energetic levels given by the high and low gate voltage.[64]

A serious problem with the McWorther noise model appears in devices with very thin gate oxides ($t_{ox} < 2$ nm). Then the tunneling time is too fast, even for traps situated close to the gate electrode/oxide interface, and no $1/f$ noise would be produced at the lowest frequencies. Instead, a roll-off in the spectrum is expected at a frequency corresponding to the tunneling time to the farthest situated traps. Low-frequency noise results have been reported for some MOSFETs with gate oxide thicknesses between 1.2 and 1.5 nm, only $1/f$ noise was observed in contradiction with the McWorther model.[45,65,66] This is an important problem that must be addressed in future noise research on ultra-scaled CMOS devices. Moreover, a trap situated in the middle of the gate oxide couples out to a flat-band voltage fluctuation with a strength given by 1/2 of the value for a trap located at the gate oxide/channel interface. Eq. (3-34) should therefore be modified by multiplying the integrand with $(1-z/t_{ox})^2$. The frequency exponent is then expected to be lower than 1.

4. MOBILITY FLUCTUATIONS

4.1 The Hooge noise model

The drain current noise generated by fluctuations in the channel carrier mobility is given according to Hooge's empirical formula

$$\frac{S_{I_D}}{I_D^2} = \frac{q\alpha_H}{fWLQ_i} \tag{3-49}$$

which is derived from Eq. (1-37) with the number of carriers in channel N replaced by WLQ_i/q. In the linear region, $Q_i = C_{ox}(V_{GS} - V_T)$, thus the normalized drain current noise depends inversely on the gate voltage overdrive. The Hooge parameter can often be considered as a constant, but the channel position under the gate oxide and the bias dependence of different scattering mechanisms likely affect the mobility fluctuation noise as discussed later. Typical values for α_H range between 10^{-3} and 10^{-6}. Values down to 10^{-7} have been observed for buried channel Si pMOSFETs,[67] and in the order of 10^{-8} for junction field-effect transistors (JFETs).[68,69]

The relation in Eq. (3-49) is only valid when the carrier density is uniform. In the saturation region, the carrier density varies parabolically along the channel and reaches zero at the drain end. Then the total channel drain current noise is evaluated by dividing the channel into small segments, each generating a noise contribution, and integrating over the channel

$$\frac{S_{I_D}}{I_D^2} = \frac{q\alpha_H}{fWL^2}\int_0^L \frac{dx}{Q_i(x)} = \{I_D = W\mu_{eff}Q_i dV/dx\} =$$
$$\frac{q\alpha_H}{fWL^2}\int_0^{V_{DS}} \frac{W\mu_{eff}}{I_D} dV = \frac{q\alpha_H \mu_{eff} V_{DS}}{fL^2 I_D}. \tag{3-50}$$

This equation is valid for all regions of operation, but V_{DS} is replaced with $V_{DS,sat}$ for $V_{DS} > V_{DS,sat}$. Using $V_{DS,sat} = (V_{GS} - V_T)/m$ and Eq. (3-8), the following expression is found for the saturated range

$$\frac{S_{I_D}}{I_D^2} = \frac{q\alpha_H \sqrt{2\mu_{eff}}}{f\sqrt{WL^3 C_{ox} m I_D}}. \tag{3-51}$$

In the subthreshold region, the drain current is dominated by diffusion

3. 1/f noise in MOSFETs

$$I_{D,diff} = \frac{WkT\mu_{eff}}{q}\frac{dQ_i(x)}{dx}. \quad (3\text{-}52)$$

Using the above expression in the integral to the left in Eq. (3-50), it is readily shown that the same final result appears. However, the drain current and the total charge density Q_i is independent on $V_{DS} \gg kT/q$. The mobility 1/f noise is also independent of V_{DS} in this case and can be written as[70]

$$\frac{S_{I_D}}{I_D^2} = \frac{\alpha_H \mu_{eff} 2kT}{fL^2 I_D}. \quad (3\text{-}53)$$

Fig. 3-11 shows a simulation of the mobility 1/f noise from subthreshold to the strong inversion regime using Eqs. (3-49) and (3-53) with a constant arbitrary α_H (other parameters for a standard pMOSFET). The mobility fluctuation noise in Fig. 3-11 is compared with the number fluctuation noise curve ($\alpha = 0$) from Fig. 3-7. From the general bias dependencies in Fig. 3-11, it can be deduced that the number fluctuation noise is most important around threshold, whereas the mobility noise is prominent both at very low currents in the subthreshold region and at very high currents in the strong inversion region.

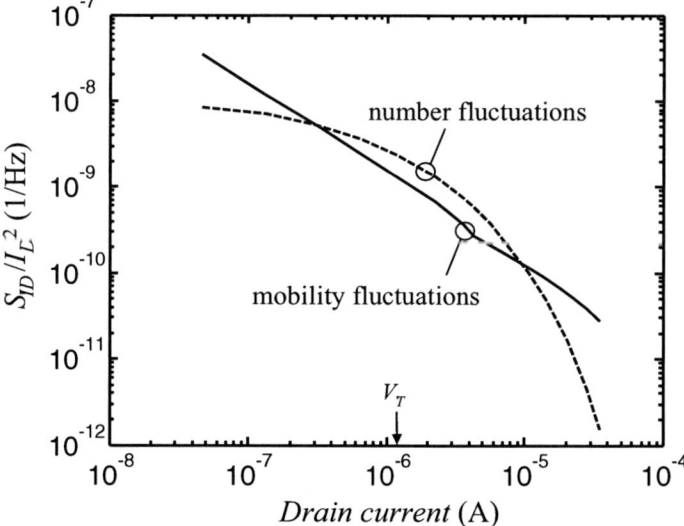

Figure 3-11. Simulation of the mobility fluctuation noise from subthreshold to the strong inversion regime using Hooge's empirical model. Eqs. (3-49) and (3-53) were used with a constant arbitrary α_H, the other parameters for a standard pMOSFET.

A popular method to perform diagnosis of the LF noise sources in MOSFETs is to compare the measured normalized drain current noise versus drain current in log-log scale with $(g_m/I_D)^2$ and $1/I_D$. This follows from Eq. (3-30) for the number fluctuations and Eq. (3-50) for the mobility fluctuations, respectively. However, note that both mechanisms may fail to accurately describe the LF noise data over the entire bias range. The mobility fluctuation noise model often provides a good fit to the noise data in strong inversion whereas the number fluctuation noise model is more useful below and around threshold.

The $1/f$ noise can be significantly higher when the current density J is inhomogeneous. In such case, Vandamme and Trefán have shown that the effective number of carriers is reduced and the $1/f$ noise is increased[71]

$$\frac{S_I}{I^2} = \frac{\alpha_H \cdot \int_\Omega J^4 d\Omega}{f \cdot n \cdot \left(\int_\Omega J^2 d\Omega\right)^2}. \tag{3-54}$$

Here, n is the carrier concentration and Ω represents the volume. As an example, current crowding at contacts can lead to deteriorated $1/f$ noise performance due to inhomogeneous current flow. Eq. (3-54) is applicable for a homogeneous conductivity but inhomogeneous current density. Inhomogeneous samples can also exhibit (much) higher $1/f$ noise, especially if they contain interfaces perpendicular to the current flow, since the noise will be generated in a smaller effective volume.

4.2   Fluctuations in phonon scattering

As mentioned in chapter 1, the mobility $1/f$ noise is suggested to be primarily generated in the phonon scattering.[72] The different scattering mechanisms that limit the channel mobility in MOSFETs depend in different ways on the vertical electric field and the density of inversion charge. Therefore, α_H is not only dependent on technology but also on the bias conditions as will be shown below. In the general case, each scattering process j generates mobility fluctuation noise with a magnitude given by the Hooge parameter of the process $\alpha_{H,j}$. If the scattering processes are independent of one another Matthiessen's rule can be applied

$$\frac{1}{\mu_{eff}} = \sum_j \frac{1}{\mu_j}. \tag{3-55}$$

The fluctuations in the different scattering processes are also assumed independent. Then we obtain

$$\frac{\Delta\mu_{eff}}{\mu_{eff}^2} = \sum_j \frac{\Delta\mu_j}{\mu_j^2} \qquad (3\text{-}56)$$

and for the power spectral densities

$$\frac{S_{I_D}}{I_D^2} = \frac{S_{\mu_{eff}}}{\mu_{eff}^2} = \sum_j \left(\frac{\mu_{eff}}{\mu_j}\right)^2 \frac{S_{\mu_j}}{\mu_j^2} = \left\{\frac{S_{\mu_j}}{\mu_j^2} = \frac{q\alpha_{H,j}}{fWLQ_i}\right\} = \sum_j \left(\frac{\mu_{eff}}{\mu_j}\right)^2 \frac{q\alpha_{H,j}}{fWLQ_i}. \qquad (3\text{-}57)$$

Thus

$$\alpha_H = \sum_j \frac{\mu_{eff}^2}{\mu_j^2} \alpha_{H,j}. \qquad (3\text{-}58)$$

If only phonon scattering generates noise

$$\alpha_H = \frac{\mu_{eff}^2}{\mu_{ph}^2} \alpha_{H,ph}. \qquad (3\text{-}59)$$

Thus, α_H varies with gate bias due to the bias dependent factor $(\mu_{eff}/\mu_{ph})^2$. The mobility ratios $(\mu_{eff}/\mu_{ph})^2$, $(\mu_{eff}/\mu_{ac})^2$, $(\mu_{eff}/\mu_{sr})^2$ and $(\mu_{eff}/\mu_{C,imp})^2$ were simulated for both electrons (open symbols) and holes (closed symbols) using the mobility model by Darwish et al.[54] and are displayed in Fig. 3-12. The phonon mobilities $\mu_{ph} = 1/(1/\mu_b + 1/\mu_{ac})$ and μ_{ac} (acoustic phonons) are more dominant for the hole transport than that for the electrons. Thus, a larger Hooge parameter can be expected for the holes. As seen in Fig. 3-12, a weak bias dependence of $\alpha_H \propto (\mu_{eff}/\mu_{ph})^2$ is predicted where α_H declines below threshold and at large inversion carrier densities. However, it is often observed from 1/f noise experiments that α_H decreases more steeply below threshold than predicted from Eq. (3-59) and increases markedly at large inversion carrier densities. In the frame of the mobility fluctuation noise model, this could either be explained by a bias dependent $\alpha_{H,ph}$ or by influence from surface roughness scattering. We will discuss these possibilities in the coming subsection.

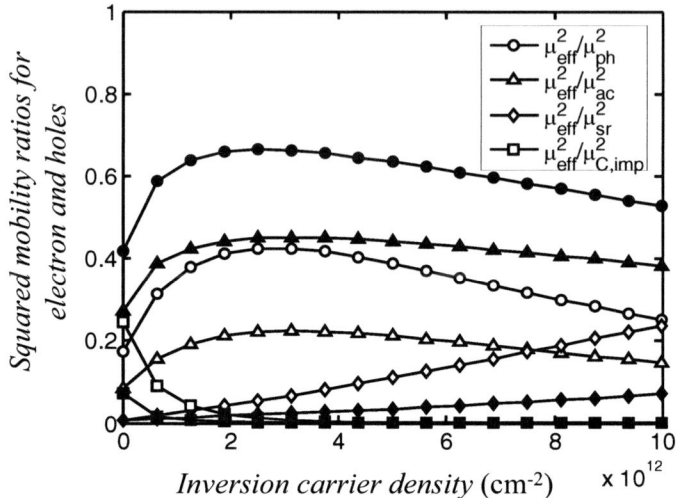

Figure 3-12. Simulation of $(\mu_{eff}/\mu_{ph})^2$, $(\mu_{eff}/\mu_{ac})^2$, $(\mu_{eff}/\mu_{sr})^2$ and $(\mu_{eff}/\mu_{C,imp})^2$ for both electrons (open symbols) and holes (closed symbols) using the mobility model by Darwish et al.[54]

4.3 Critical discussion

The Hooge model for mobility fluctuation noise has shown a remarkable success in describing the $1/f$ noise in homogeneous semiconductor samples and metals. However, the lack of a physics based theoretical model is a weakness, and the model has been criticized from fundamental physics viewpoints by Weissman among others.[73] In MOSFETs, the $1/f$ noise can be generated by number fluctuations due to trapping and detrapping of carriers in the gate oxide. Nowadays, this school of thought dominates as the primary explanation of $1/f$ noise in such devices. Still, the Hooge noise model satisfactorily describes the $1/f$ noise in pMOSFETs, but a gate voltage dependent Hooge parameter, $\alpha_H \propto 1/V_{GT}$, must sometimes be assumed in the description of nMOSFETs. However, one unsolved problem with the Hooge noise model is that it is less accurate in the subthreshold region. It is also found that the $1/f$ noise is sensitive to the crystalline quality, channel position and vertical electric field. We will discuss these observations in terms of the mobility fluctuation noise model and finally present a possible qualitative explanation of the mobility noise.

4.3.1 Subthreshold region

For both nMOSFETs and pMOSFETs, a roll-off in the S_I/I^2 curve at gate biases below the threshold voltage is often observed (see Fig. 3-7). In such cases, the apparent Hooge parameter decreases with decreasing current in the subthreshold region according to Eq. (3-50). From a modeling point of view, any bias dependence can be included in α_H, just like N_t was suggested to vary with energy in order to describe the $1/f$ noise in pMOSFETs. However, a good model should be physics based, be able to accurately model the observed behaviour and predict effects related to technology. Our approach in the following is to review and discuss refinements of the Hooge noise model starting from the basic assumptions of the model.

One of the early improvements of the Hooge noise model was the finding that different scattering mechanisms contribute with different magnitudes to the fluctuations in the mobility (see Eq. 3-58) and that the phonon scattering was proposed as the dominant source of the $1/f$ noise. A somewhat lower α_H in subthreshold can therefore be expected due to a stronger influence of Coulomb scattering which dilutes the noise in the phonon scattering. Vandamme and Vandamme also proposed that the number of carriers involved in the fluctuation process cannot go below a constant value determined by the thermal voltage.[74] Then N in Eq. (1-37) for a MOSFET is replaced by $N + N_0$, where $N_0 = WLC_{ox}kT/q^2$ is a constant. However, the theoretical basis for this assumption is still dubious.

4.3.2 Crystalline quality

It has been observed that the mobility fluctuation noise is related to the crystalline quality of the sample.[75] For example, the presence of lattice defects created by proton or ion irradiation causes increased $1/f$ noise. If the damaged material is annealed, the $1/f$ noise reduces and may approach the original value. Vandamme and Oosterhoff found a $1/f$ noise reduction in ion implanted Si resistors by a factor of at least 50 after annealing at $T \geq 750$ °C.[76] The carrier mobility, on the other hand, only varied between 360 cm^2/Vs and 270 cm^2/Vs. Moreover, Gaubert et al. recently reported that the RMS value of the surface roughness could be related to the $1/f$ noise performance.[77] They observed a pronounced improvement of the $1/f$ noise by a factor of 100, achieved by an improved cleaning and gate oxidation process, which primarily was linked to a reduced microroughness of the interface. Thus, the mobility fluctuation noise mechanism seems more sensitive to the mobility than predicted by Eq. (3-59). What can be the physical origin of the crystalline quality dependence of the $1/f$ noise and what are the consequences?

4.3.3 1/f noise dependence on channel position

Field-effect transistors with a buried channel for the current transport can often show significantly lower $1/f$ noise than that for surface channel devices. This has been reported for buried channel Si MOSFETs,[67,78-80] SiGe pMOSFETs,[12,81] SOI four-gate transistors,[82] and JFETs[68,69] to mention some examples. The principle of the JFET is that the gate voltage modulates the width of the depletion region, which narrows or widens the conduction channel in the un-depleted part of the semiconductor. Thus, if the conduction channel is separated from any oxide interface, lower $1/f$ noise can be obtained. These results are in favour of the number fluctuation theory, since the oxide traps cannot generate $1/f$ noise if the carriers are isolated from them. But what about the $1/f$ noise in the buried channel devices if the number fluctuation noise is eliminated? The $1/f$ noise does not disappear completely since the mobility fluctuation noise still remains. However, is it possible that the mobility fluctuation noise is enhanced by the gate oxide interface as well? The mobility is clearly reduced for surface conduction due to additional surface scattering (acoustic phonons and surface roughness). Moreover, experimental evidence suggests that the Hooge parameter is sensitive to the crystalline quality, which is deteriorated close to the interface. Therefore, another possibility to explain the higher $1/f$ noise when the carriers are in close proximity to the gate oxide surface is by increased mobility fluctuation noise.

In the authors' work it has been suggested that the mobility fluctuation noise depends on the distance of the carriers from the gate oxide interface. The conduction path of the carriers can be modified by device engineering and the operating conditions. For example, a reverse bias on the substrate-source junction of a bulk MOSFET increases the vertical electric field and the carriers are pushed closer towards the gate oxide. Fig. 3-13 shows how the Hooge parameter increases rapidly when the average carrier-oxide distance is about 2 nm. The behaviour in Fig. 3-13 was obtained for both SOI and bulk Si pMOSFETs and for different bias conditions, indicating a general trend. Thus, the mobility fluctuation $1/f$ noise also depends on the location of the conduction path due to the influence of the gate oxide/Si interface.

3. 1/f noise in MOSFETs

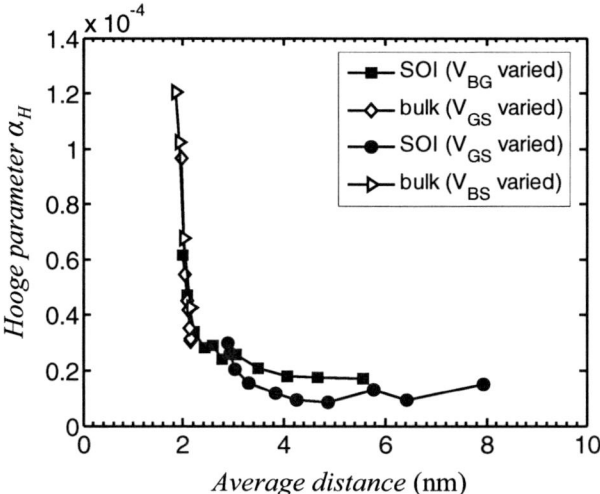

Figure 3-13. The Hooge parameter, extracted from LF noise measurements on SOI and bulk Si pMOSFET, plotted vs. average distance of the inversion carriers from the gate oxide interface. The carrier-oxide distance was obtained from TCAD simulations.

4.3.4 Source of mobility fluctuations

In terms of a physical explanation and modeling of the dependence on carrier-oxide distance and crystal quality, two main sources can be identified. The variations in α_H originates (i) from the noise in the phonon scattering but with a non-constant $\alpha_{H,ph}$, or (ii) noise in the surface roughness scattering that takes over as the dominant mobility fluctuation noise mechanisms as was suggested by the authors.[12] The mobility fluctuates since the carriers' position and velocity are random. The surface is fixed but the carriers are not, which lead to fluctuations in the surface roughness scattering as well as other scattering processes. The effective mobility μ_{eff} can be written as

$$\mu_{eff}^{-1} = \mu_{sr}^{-1} + \mu_{a}^{-1} \tag{3-60}$$

where μ_{sr} is the mobility limited by surface roughness scattering, and μ_a here represents the mobility limited by scattering mechanisms other than surface roughness scattering. According to Eq. (3-58) the following relationship for the Hooge parameter can be derived

$$\alpha_H = \mu_{eff}^2 \left(\frac{\alpha_{H,sr}}{\mu_{sr}^2} + \frac{\alpha_{H,a}}{\mu_a^2} \right) \qquad (3\text{-}61)$$

where $\alpha_{H,sr}$ and $\alpha_{H,a}$ are the Hooge parameters for the corresponding scattering processes, which are considered as constants here. Fig. 3-14 shows a simulation of the $1/f$ noise in a Si pMOSFET using Eq. (3-61) with $\alpha_{H,sr} = 1.56\times10^{-3}$ and $\alpha_{H,a} = 1.1\times10^{-5}$. The $1/f$ noise was measured as a function of the gate and substrate voltages. As seen, the simulation shows a good agreement with the measured data for both the V_{GT} and V_{BS} dependencies. The discrepancy for V_{GT} above 0.5 V is, at least partly, due to influence from noise generated in the source and drain resistances. The S_I/I^2 curve tends to flatten out and show a sublinear $1/I_D$ dependence at high gate voltage overdrives V_{GT}, as seen in this experiment. A similar behaviour has been reported in the literature by other groups as well.[83,84] It was found in this case that α_H for the Si pMOSFET increased with V_{GT} as

$$\alpha_H \propto V_{GT}^{0.38} \qquad (3\text{-}62)$$

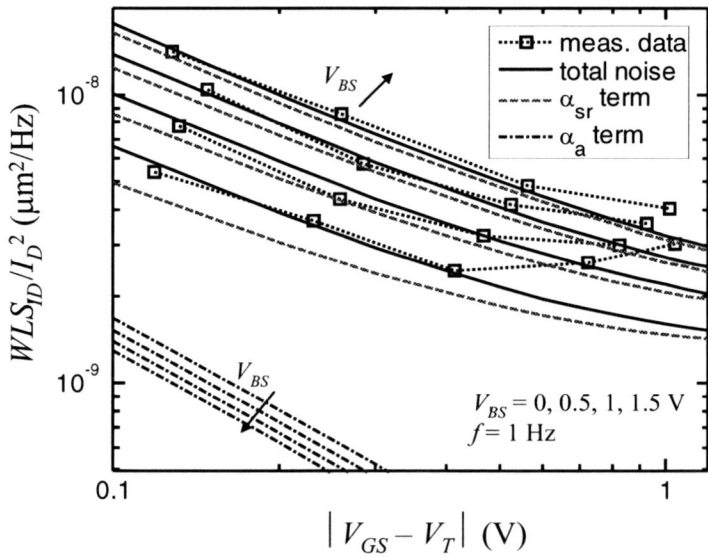

Figure 3-14. Simulated (lines) and measured (open squares) drain current noise on a Si pMOSFET for different V_{BS}.[12] Eq. (3-61) was used for α_H. The total noise from the simulation is shown as well as the contributions from the $\alpha_{H,sr}$ term and $\alpha_{H,a}$ term in Eq. (3-61).

3. 1/f noise in MOSFETs

The mobility ratio $(\mu_{eff}/\mu_{sr})^2$ increases with increasing inversion carrier density (increasing vertical electric field), as found in Fig. 3-12, which explains the observation in Eq. (3-62). This phenomenon can also be linked to the effects of the substrate bias (see next section) as the electric field increases with increasing gate voltage overdrive as well as with increasing substrate bias. Thus, if the 1/f noise in the surface roughness scattering is dominant, the dependence on gate and substrate bias as well as on channel position could be explained. This exercise shows that by invoking surface roughness scattering it is possible to improve the *modeling* of the 1/f noise.

It was suggested by Melkonyan *et al.* that the mobility fluctuation 1/f noise is related to the phonon-phonon scattering rate.[85] The following relationship was derived

$$\alpha_H = \frac{K}{\tau_{ph-ph}} \qquad (3\text{-}63)$$

where K is a constant and $1/\tau_{ph-ph}$ is the rate by which phonons scatter against other phonons. Let us generalize Eq. (3-63) somewhat and assume that the Hooge parameter is proportional to the overall phonon scattering rate. The phonon mean free path is strongly sensitive to lattice imperfections, defects, surfaces etc.[86] For example, the thermal conductivity, which is proportional to the phonon mean free path, was found to increase by a factor of ~500 in the experiment by Takabatake and co-workers when a heavily ion implanted semiconductor specimen was annealed at 750 °C.[87] Therefore, the following qualitative explanation of the mobility fluctuation noise is suggested. Phonons scatter randomly which introduce random fluctuations in the phonon-electron scattering as well as the in the electron distribution, giving rise to 1/f fluctuations in the mobility. The more the phonons scatter (shorter mean free path) the stronger the mobility fluctuations.

One important conclusion here is that the 1/f noise, both due to number and mobility fluctuations, is sensitive to the gate oxide/channel interface properties and the current transport close to it. Parameters such as vertical electric field, scattering mechanisms, position of the inversion carriers should be carefully investigated and included in the noise 1/f models. The exact origin of the mobility fluctuations is still not fully established. However, interesting theoretical models have been reported recently that have shed some light over the phenomenon (see also chapter 1). Here, we have tried to take these theories one step further in order to understand the 1/f noise in MOSFETs.

5. IMPACT OF SUBSTRATE VOLTAGE

The $1/f$ noise is often found to decrease when a voltage V_{BS} is applied to the bulk (substrate) terminal forward biasing the substrate/source junction,[13,14,28,45] which may be exploited in circuits to reduce the noise figure or phase noise. Conversely, the $1/f$ noise increases when the substrate/source junction is reverse biased.[12] Another related observation is that dynamic threshold (DT) MOSFETs, where the bulk terminal is connected to the gate $V_B = V_G$, can present higher transconductance and significantly reduced $1/f$ noise than that when the device is operated in the conventional mode with the bulk grounded.[88-90] Deen and Marinov found a reduction in the $1/f$ noise magnitude by 8 dB/V when a forward bias was applied on the bulk of a Si pMOSFET.[14] In the authors' work, a corresponding reduction around 5 dB/V has been observed in Si pMOSFETs with high-k gate dielectrics.[28] These results are discussed below and compared with other results published in literature. First, we will explain the effect of V_{BS} on the static device parameters (body effect).

The width of the depletion region is changed due to the voltage V_{BS}. The depletion charge increases when the substrate is reverse biased (V_{BS} positive for pMOS, negative for nMOS), and decreases when the substrate is forward biased

$$Q_d = \sqrt{2\varepsilon_{si} q N_{sub}(2\psi_B + V_{BS})}. \tag{3-64}$$

The threshold voltage depends on the depletion charge and is therefore shifted by the voltage on the bulk terminal, $|V_T|$ is increased by a reverse bias and decreased by a forward bias. Eq. (3-3) must be modified for a nonzero V_{BS} according to

$$V_T = V_{fb} \pm 2\psi_B \pm \frac{\sqrt{2\varepsilon_{si} q N_{sub}(2\psi_B + V_{BS})}}{C_{ox}}. \tag{3-65}$$

The effective mobility varies with the effective electric field as discussed in section 2.1.2. The effective electric field is given as

$$E_{eff} = \frac{1}{\varepsilon_{Si}}(Q_d + \eta Q_i) \tag{3-66}$$

with η usually taken as 1/2 for electrons and 1/3 for holes, respectively. The effective electric field is increased by a reverse substrate bias (and increasing gate voltage overdrive), which results in a decrease in the mobility. The opposite occurs for a forward substrate bias. Fig. 3-15 shows g_m plotted versus V_{GS} for a Si pMOSFET with HfAlO$_x$ gate dielectrics demonstrating both the effect on V_T and on μ_{eff}. The average inversion carrier distance from the gate oxide interface increases for a forward substrate bias, vice versa for a reverse bias, as evidenced from the simulation in Fig. 3-16.

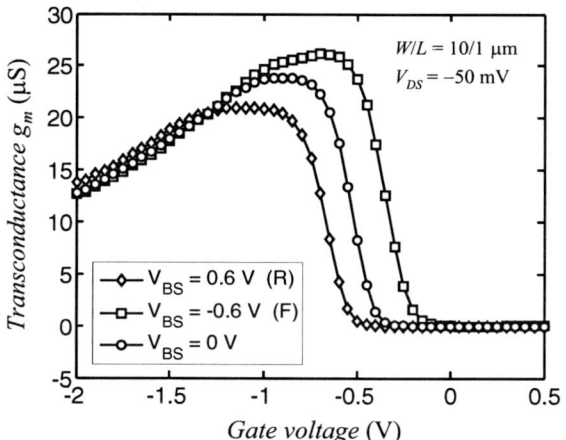

Figure 3-15. Transconductance plotted as a function of gate voltage for a Si pMOSFET with HfAlO$_x$ gate dielectrics. The substrate was biased with $V_{BS} = 0$, -0.6 and 0.6 V.

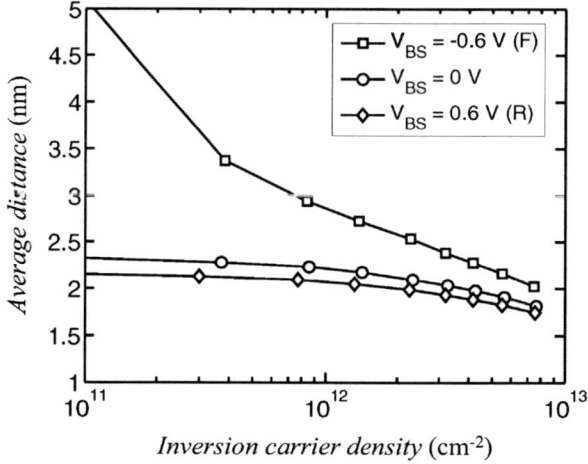

Figure 3-16. TCAD simulation of the average distance of the inversion charge from the oxide interface for different V_{BS}. $t_{ox} = 2.8$ nm and $N_d = \sim 5 \times 10^{17}$ cm^{-3} were used in the simulations.

The 1/*f* noise has been found to decrease with forward substrate bias and increase with a reverse bias in a large number of studied devices including Si pMOSFETs, buried SiGe channel pMOSFETs and pMOSFETs with high-k gate dielectrics. The substrate bias effect is illustrated for the abovementioned devices in Fig. 3-17. The influence on the substrate voltage is weak in some cases (as in Fig. 3-17(b)), but still discernable.

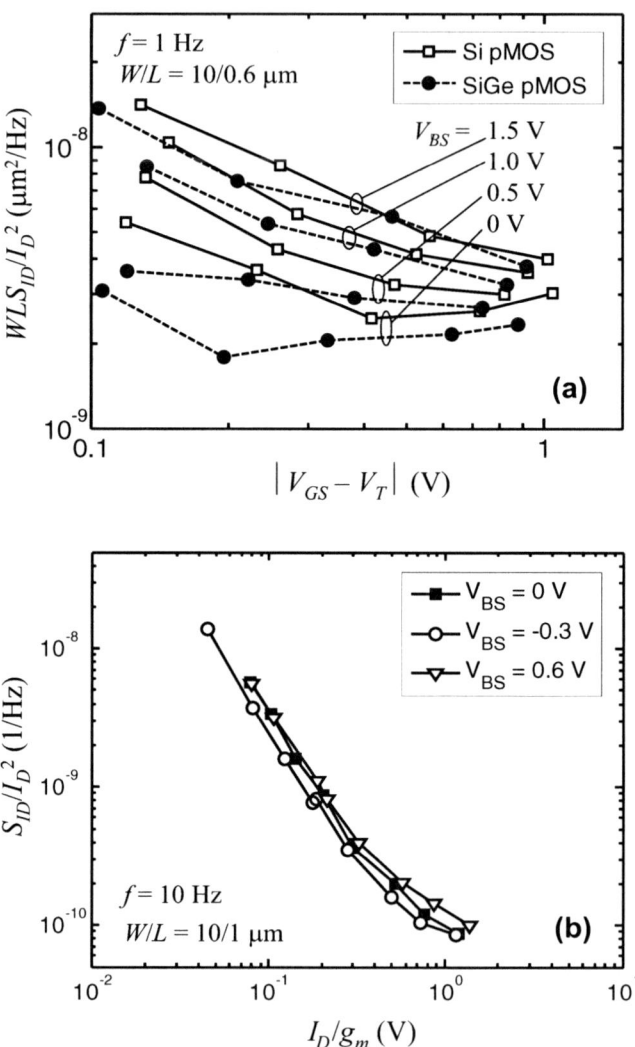

Figure 3-17. Drain current noise measured at different V_{BS} on (a) a Si and a buried SiGe (6-nm Si-cap) channel pMOSFET with a poly-Si/SiO$_2$ gate stack (reprinted from von Haartman et al.[12] with permission from IEEE) and (b) a surface SiGe channel pMOSFET with a poly-SiGe/HfO$_2$/Al$_2$O$_3$ gate stack.

3. 1/f noise in MOSFETs

The physical mechanism behind the noise lowering has been studied thoroughly. In the literature, several suggestions have been presented to explain the effect of the substrate bias. Park et al. examined nMOS transistors and found a noise reduction by one order of magnitude in weak inversion, but an almost independent behaviour in strong inversion.[13] A similar observation was made by Marin et al. on MOSFETs from a 130 nm CMOS technology, ~50% lower noise was found in weak inversion in their work.[45] These results can be explained by the fact that the depletion capacitance C_d changes with the substrate voltage (increases with forward substrate bias, opposite for reverse bias), which affects the number fluctuation noise according to Eq. (3-40). However, the results shown in Fig. 3-17 suggest a strong influence (also) above threshold. Ahsan and Schroder go a step further and claim that the correlated mobility fluctuations also are affected by the substrate bias.[91] This is explained by the fact that the Coulomb interaction depends on the distance between the oxide charge and the channel carriers, which is modulated by the substrate bias as discussed above. However, the effective mobility is also affected by the substrate bias. It has been shown that the factor $\alpha \cdot \mu_{eff}$ determining the strength of the correlated mobility fluctuations (see Eq. 3-30) rather would change with the substrate voltage in the opposite direction than the $1/f$ noise does.[92] The correlated mobility fluctuation noise should therefore have a negligibly small influence on the $1/f$ noise variations with the substrate bias.

If we consider that the $1/f$ noise depends on the vertical electric field and/or channel position, we can explain the $1/f$ noise variation with substrate bias in *strong inversion* due to mobility fluctuation noise according to the behaviour in Fig. 3-13. An alternative description is illustrated in Fig. 3-18 where α_H is studied versus the effective vertical electric field for three different pMOSFETs with high-k gate dielectrics.

Finally, the quantum effects induced by the high vertical electrical field have been investigated by Mercha et al.[93] An empirical correction factor σ_{2D}/σ_{3D}, which increases strongly with the vertical electric field, was proposed and included in Eq. (3-38). This effect can also potentially explain the noise behaviour on V_{BS}.

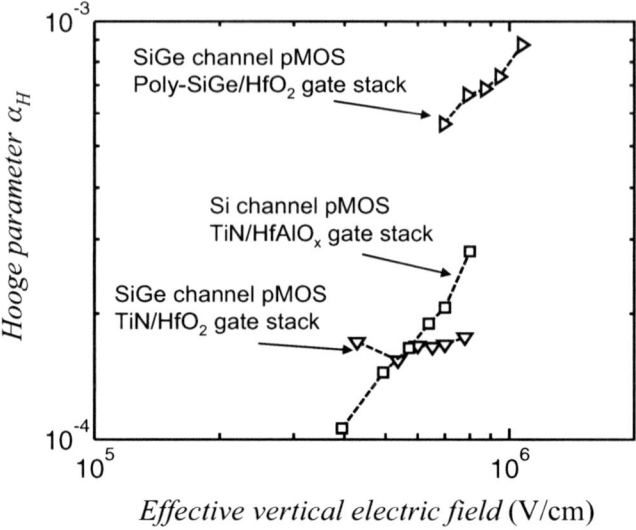

Figure 3-18. The Hooge parameter plotted vs. effective vertical electric field for Si and SiGe pMOSFETs with high-k gate dielectrics.[28] The electric field was varied by a voltage on the bulk terminal V_{BS} in the range −0.6 to 1.2 V. The Hooge parameter was extracted in strong inversion regime.

6. COMPACT NOISE MODELS

A compact model is a mathematical description of a semiconductor device used for circuit simulation and technology development. It is of crucial importance to be able to perform reliable and accurate simulations of the circuit behaviour when designing electronic circuits. For simulation of $1/f$ noise, the Berkeley short channel IGFET (BSIM3) model has received a wide acceptance. This model was originally proposed by Hung *et al.* and combines number fluctuations and correlated mobility fluctuations in one model by using three basic noise parameters (*NOIA*, *NOIB*, *NOIC*).[31] It should be mentioned that a slightly modified version of the BSIM3 $1/f$ noise model has been implemented in other compact models also, like the Philips MOS model 11. We are going to present and discuss the $1/f$ noise models in the BSIM3 version 3 model package (abbreviated BSIM3v3), basically the same noise equations are also used in the extended BSIM4 model.

Two $1/f$ noise models are implemented in BSIM3v3, one is a simple SPICE2 noise model and the other one is the BSIM3 noise model. The SPICE2 noise model is given below

3. 1/f noise in MOSFETs

$$S_{I_D} = \frac{KF \cdot I_D^{af}}{C_{ox} L_{eff}^2 f^{ef}}. \tag{3-67}$$

For $af = 1$, this model is identical to the number fluctuation noise model in the saturation region, see Eq. (3-30) with $\alpha = 0$. The SPICE2 model is good for hand calculations but is too simple for accurate simulations. The BSIM3 model includes the influence of correlated mobility fluctuations via the parameters *NOIB* and *NOIC*. For $V_{GS} > V_T + 0.1$ V, the BSIM3 model is written according to[94]

$$\begin{aligned} S_{I_D} = &\frac{kTq^2 \mu_{eff} I_D}{C_{ox}(L_{eff} - 2 \cdot LINTNOI)^2 A_{bulk} f^{ef} \cdot 10^8} \times \\ &\left[NOIA \ln\left(\frac{N_0 + N^*}{N_l + N^*}\right) + NOIB(N_0 - N_l) + \frac{NOIC}{2}(N_0^2 - N_l^2) \right] + \\ &\frac{kT I_D^2 \Delta L_{clm} \cdot 10^{-8}}{W_{eff}(L_{eff} - 2 \cdot LINTNOI)^2 f^{ef}} \frac{NOIA + NOIB \cdot N_l + NOIC \cdot N_l^2}{(N_l + N^*)^2}. \end{aligned} \tag{3-68}$$

The parameters N_0 and N_l are the charge densities at the source and drain side, respectively

$$N_0 = C_{ox}(V_{GS} - V_T)/q \tag{3-69}$$

$$N_l = C_{ox}\left[V_{GS} - V_T - \min(V_{DS}, V_{DS,sat})\right]/q. \tag{3-70}$$

ΔL_{clm} is the channel length reduction due to channel length modulation effect. The parameter A_{bulk} describes the bulk charge effect and corresponds to the coefficient m used here. The parameter N^* is equal to 2×10^{14} m^{-2} in the BSIM3v3 model but is evaluated as

$$N^* = kT(C_{ox} + C_d + CIT)/q^2 \tag{3-71}$$

in the BSIM4 model, where C_d is the depletion capacitance and *CIT* is a model parameter. For $V_{GS} < V_T + 0.1$ V, the 1/f noise is given as

$$S_{I_D} = \frac{S_{\lim} S_{wi}}{S_{\lim} + S_{wi}} \qquad (3\text{-}72)$$

where $S_{\lim}$ is the $1/f$ noise calculated at $V_{GS} = V_T + 0.1$ V and

$$S_{wi} = \frac{kT \cdot I_D^2 \cdot NOIA}{W_{eff} L_{eff} f^{ef} N^{*2} \cdot 10^8} \qquad (3\text{-}73)$$

is the $1/f$ noise in subthreshold. The expression in Eq. (3-72) ensures continuity between the $1/f$ noise equations in the two regions of operation. The parameters used in the noise models are listed in Table 3-2 along with their default values.

Table 3-2. Noise parameters in the BSIM3v3 model.

Parameter name	Description	Default value	Unit
NOIA	BSIM3 Noise parameter A	1×10^{20} (nMOS)	$m^{-3} eV^{-1}$
		9.9×10^{18} (pMOS)	
NOIB	BSIM3 Noise parameter B	5×10^4 (nMOS)	$m^{-1} eV^{-1}$
		2.4×10^3 (pMOS)	
NOIC	BSIM3 Noise parameter C	-1.4×10^{-12} (nMOS)	$m \cdot eV^{-1}$
		1.4×10^{-12} (pMOS)	
af	SPICE2 Current exponent	1	-
ef	Frequency exponent	1	-
KF	SPICE2 Noise coefficient	0	
LINTNOI	Length reduction parameter offset	0	m

The BSIM3 model matches the number fluctuation noise model in Eq. (3-30) with

$$\begin{aligned} NOIA &\cong N_t \\ NOIB &\cong \alpha \mu_{eff} q N_t \\ NOIC &\cong \alpha^2 \mu_{eff}^2 q^2 N_t / 2 \end{aligned} \qquad (3\text{-}74)$$

for the saturation region. Obviously, the noise parameters are not independent since $NOIC = NOIB^2/2NOIA$. Note the factor 10^8 in Eq. (3-68) correspond to $1/\lambda$ in Eq. (3-38). $\lambda \approx 0.1$ nm ($1/\lambda \approx 10^8$ cm^{-1}) for the SiO$_2$/Si system. However, if the distance is expressed in meter (standard unit in BSIM3) instead of centimeter, the right hand sides of the expressions in Eq. (3-74) must be divided by a factor 100 to be correct. This unit confusion has been corrected in the BSIM4 model since the factor 10^8 is replaced by 10^{10}.

The BSIM3 noise model can be used to model the Hooge mobility fluctuation noise also. By comparing Eq. (3-68) with the Hooge noise model

3. 1/f noise in MOSFETs   95

for the saturation region given by Eq. (3-51), it is found that *NOIA = NOIC = 0* and *NOIB* = $10^8 \cdot \alpha_H/kT$. The BSIM3 model was originally developed from the number fluctuation noise theory but is using three noise parameters instead of two that would be enough according to Eq. (3-30). The three noise parameters give flexibility to the model without making it too complex, the BSIM3 1/f noise model has therefore been successful in circuit simulations and can describe most observed 1/f noise behaviours.

7. INPUT REFERRED NOISE

We have so far discussed the 1/f noise in the drain current. In analogy with the amplifier noise model, the 1/f noise in the drain current can be referred to an *equivalent* input gate voltage source. The noise is in fact not generated from gate voltage fluctuations because the gate terminal is set to a fixed voltage. In case of number fluctuations, the noise is generated by fluctuations in the oxide charge which cause fluctuations in the surface potential. The mobility fluctuation noise stems from the conduction path of the carriers. The equivalent input gate voltage noise is therefore only a mathematical construction, which is calculated from the drain current noise as

$$S_{V_G} = \frac{S_{I_D}}{g_m^2}. \tag{3-75}$$

Note that both the number and mobility fluctuations are generated without the drain current being present, but the fluctuations in the oxide charge and mobility are sensed in the drain current. The equivalent gate voltage noise due to the number fluctuation noise model is found by using Eqs. (3-30) and (3-38)

$$S_{V_G} = \frac{q^2 kT\lambda N_t}{f^\gamma WLC_{ox}^2}\left(1 + \frac{\alpha\mu_{eff}C_{ox}I_D}{g_m}\right)^2. \tag{3-76}$$

The gate voltage noise due to mobility fluctuation noise is given in the linear regime as

$$S_{V_G} = \frac{q\alpha_H}{fWLC_{ox}}(V_{GS} - V_T)[1 + \theta(V_{GS} - V_T)]^2. \tag{3-77}$$

Here, θ is a mobility coefficient used to describe the mobility attenuation with increasing gate voltage overdrive according to

$$\mu_{eff} = \frac{\mu_0}{1+\theta(V_{GS}-V_T)} \qquad (3\text{-}78)$$

where μ_0 is the low-field mobility. Thus, the number fluctuations give a constant gate voltage noise (for small α), whereas the gate voltage noise increases proportional to the gate voltage overdrive when mobility fluctuations dominate.

SUMMARY

- The equivalent noise circuit (Fig. 3-5) of the MOSFET was presented and the output noise generated from the channel and S/D regions was calculated (Eqs. 3-12 and 3-13).
- The two sources of $1/f$ noise, number fluctuations and mobility fluctuations, were discussed and expressions for the drain current noise PSD were derived.
  Eq. (3-30) and Eq. (3-38) for number fluctuations:

$$S_{I_D} = \frac{q^2 kT\lambda N_t}{f^\gamma WLC_{ox}^2}\left(1+\frac{\alpha\mu_{eff}C_{ox}I_D}{g_m}\right)^2 g_m^2 \qquad (3\text{-}79)$$

This expression is derived for the McWorther model, which assumes that the transitions to and from the traps in the gate oxide occur by tunneling. The mobility fluctuation noise model is written as (Eqs. 3-49 and 3-50):

$$S_{I_D} = \frac{q\alpha_H \mu_{eff} V_{DS} I_D}{fL^2} \quad (= \frac{q\alpha_H I_D^2}{fWLC_{ox}(V_{GS}-V_T)} \text{ in the linear region)} \qquad (3\text{-}80)$$

- The $1/f$ noise is found to decrease for a forward bias on the body-source junction (forward substrate bias), and vice versa for a reverse bias. This phenomenon is not yet described by the standard $1/f$ noise models.
- The SPICE2 (parameters *KF*, *af*) and BSIM3 (noise parameters *NOIA*, *NOIB* and *NOIC*) compact $1/f$ noise models were discussed.

REFERENCES

1. G. E. Moore, Cramming more components onto integrated circuits, Electronics **38** (1965).
2. P. H. Woerlee, M. J. Knitel, R. van Langevelde, D. B. M. Klaassen, L. F. Tiemeijer, A. J. Scholten, and A. T. A. Zegers-van Duijnhoven, RF-CMOS performance trends, *IEEE Trans. Electron Devices* **48**, 1776-1782 (2001).
3. A. A. Abidi, RF CMOS comes of age, in *Proc. Symp. VLSI Circuits*, 2003, pp. 113-116.
4. J. L. Liou and F. Schwierz, RF MOSFET: recent advances, current status and future trends, *Solid-State Electron.* **47**, 1881-1895 (2003).
5. H. S. Bennett, R. Brederlow, J. C. Costa, P. E. Cottrell, W. M. Huang, A. A. Immorlica, J.-E. Mueller, M. Racanelli, H. Shichijo, C. E. Weitzel, and B. Zhao, Device and technology evolution for Si-based RF integrated circuits, *IEEE Trans. Electron Devices* **52**, 1235-1258 (2005).
6. A. Mercha, W. Jeamsaksiri, J. Ramos, D. Linten, S. Jenei, P. Wambacq, and S. Decoutere, Impact of scaling on analog/RF performance, in *Proc. IEEE Int. Conf. Solid-State and Integrated Circuits Technology*, 2005, pp. 147-152.
7. K. Lee, I. Nam, I. Kwon, J. Gil, K. Han, S. Park, B.-I. Seo, The impact of semiconductor technology scaling on CMOS RF and digital circuits for wireless application, *IEEE Trans. Electron Devices* **52**, 1415-1422 (2005).
8. E. Simoen, and C. Claeys, On the flicker noise in submicron silicon MOSFETs, *Solid-State Electron.* **43**, 865-882 (1999).
9. G. Ghibaudo, and T. Boutchacha, Electrical noise and RTS fluctuations in advanced CMOS devices, *Microelectron. Reliab.* **42**, 573-582 (2002).
10. L. K. J. Vandamme, X. Li, and D. Rigaud, $1/f$ noise in MOS devices, mobility or number fluctuations?, *IEEE Trans. Electron Devices* **41**, 1936-1945 (1994).
11. J. Chang, A. A. Abidi, and C. R. Viswanathan, Flicker noise in CMOS transistors from subthreshold to strong inversion at various temperatures, *IEEE Trans. Electron Devices* **41**, 1965-1971 (1994).
12. M. von Haartman, A.-C. Lindgren, P.-E. Hellström, B. G. Malm, S.-L. Zhang, and M. Östling, $1/f$ noise in Si and $Si_{0.7}Ge_{0.3}$ pMOSFETs, *IEEE Trans. Electron Devices* **50**, 2513-2519 (2003).
13. N. Park and K. K. O, Body bias dependence of $1/f$ noise in NMOS transistors from deep-subthreshold to strong inversion, *IEEE Trans. Electron Devices* **48**, 999-1001 (2001).
14. M. J. Deen and O. Marinov, Effect of forward and reverse substrate biasing on low-frequency noise in silicon PMOSFETs, *IEEE Trans. Electron Devices* **49**, 409-413 (2002).
15. Y. Taur and T. H. Ning, *Fundamentals of modern VLSI devices* (Cambridge University Press, Cambridge, 1998)
16. M. Lundstrom, *Fundamentals of carrier transport* (Cambridge University Press, Cambridge, 2000).
17. S. Takagi, A. Toriumi, M. Iwase, and H. Tango, On the universality of inversion layer mobility in Si MOSFET's: part I-effects of substrate impurity concentration, *IEEE Trans. Electron Devices* **41**, 2357-2362 (1994).
18. S. A. Schwarz and S. E. Russek, Semi-empirical equations for electron velocity in silicon: part II-MOS inversion layer, *IEEE Trans. Electron Devices* **ED-30**, 1634-1639 (1983).
19. V. M. Agostinelli, Jr., H. Shin, and A. F. Tasch, Jr., A comprehensive model for inversion layer hole mobility for simulation of submicrometer MOSFET's, *IEEE Trans. Electron Devices* **38**, 151-159 (1991).
20. F. Gámiz, J. B. Roldán J. E. Carceller, and P. Cartujo, Monte Carlo simulation of remote-Coulomb-scattering-limited mobility in metal-oxide-semiconductor transistors, *Appl. Phys. Lett.* **82**, 3251-3253, (2003).

21. S. M. Sze, *Semiconductor devices, physics and technology* (John Wiley & Sons, New York, 1985).
22. Y. Tsividis, *Operation and modeling of the MOS transistor* (WCB McGraw-Hill, Boston, 1999).
23. A. J. Scholten, L. F. Tiemeijer, R. van Langevelde, R. J. Havens, A. T. A. Zegers-van Duijnhoven, and V. C. Venezia, Noise modeling for RF CMOS circuit simulation, *IEEE Trans. Electron Devices* **50**, 618-632 (2003).
24. A. A. Abidi, High-frequency noise measurements on FET's with small dimensions, *IEEE Trans. Electron Devices* **ED-33**, 1801-1805 (1986).
25. P. Klein, An analytical thermal noise model of deep submicron MOSFET's, *IEEE Electron Device Lett.* **20**, 399-401 (1999).
26. R. Brederlow, G. Wenig, and R Thewes, Investigation of the thermal noise of MOS transistors under analog and RF operating conditions, in *Proc. European Solid-State Device Research Conf. (ESSDERC)*, 2002, pp. 87-90.
27. A. L. McWorther, *Semiconductor surface physics* (University of Pennsylvania Press, Philadelphia, 1957).
28. M. von Haartman, B. G. Malm, and M. Östling, Comprehensive study on low-frequency noise and mobility in Si and SiGe pMOSFETs with high-κ gate dielectrics and TiN gate, *IEEE Trans. Electron Devices* **53**, 836-843 (2006).
29. Y. Akue Allogo, M. de Murcia, J. C. Vildeuil, M. Valenza, P. Llinares, and D. Cottin, $1/f$ noise measurements in n-channel MOSFETs processed on 0.25 μm technology Extraction of BSIM3v3 parameters, *Solid-State Electron.* **46**, 361-366 (2002).
30. F. Crupi, P. Srinivasan, P. Magnone, E. Simoen, C. Pace, D. Misra, and C. Claeys, Impact of the interfacial layer on the low-frequency noise (1/f) behaviour of MOSFETs with advanced gate stacks, *IEEE Electron Device Lett.* **27**, 688-691 (2006).
31. K. K. Hung, P. K. Ko, C. Hu, and Y. C. Cheng, A unified model for the flicker noise in metal-oxide-semiconductor field-effect transistors, *IEEE Trans. Electron Devices* **37**, 654-665 (1990).
32. E. P. Vandamme and L. K. J. Vandamme, Critical discussion on unified $1/f$ noise models for MOSFETs, *IEEE Trans. Electron Devices* **47**, 2146-2152 (2000).
33. G. Ghibaudo, O. Roux, Ch. Nguyen-Duc, F. Balestra, and J. Brini, Improved analysis of low-frequency noise in field-effect MOS transistors, *Phys. Stat. Sol. A* **124**, 571-581 (1991).
34. S. Christensson, I. Lundström, and C. Svensson, Low-frequency noise in MOS transistors-I theory, *Solid-State Electron.* **11**, 797-812 (1968).
35. R. Jayaraman and C. G. Sodini, A $1/f$ noise technique to extract the oxide trap density near the conduction band edge of silicon, *IEEE Trans. Electron Devices* **36**, 1773-1782 (1989).
36. N. Lukyanchikova, M. Petrichuk, N. Garbar, E. Simoen, A. Mercha, C. Claeys, H. van Meer, and K. De Meyer, The $1/f^{1.7}$ noise in submicron SOI MOSFETs with 2.5 nm nitrided gate oxide, *IEEE Trans. Electron Devices* **49**, 2367-2370 (2002).
37. G. Reimbold, Modified $1/f$ trapping noise theory and experiments in MOS transistors biased from weak to strong inversion-influence of interface states, *IEEE Trans. Electron Devices* **ED-31**, 1190-1198 (1984).
38. H.-S. Fu and C.-T. Sah, Theory and experiments on surface $1/f$ noise, *IEEE Trans. Electron Devices* **ED-19**, 273-285 (1972).
39. C. Surya and T. Y. Hsiang, A thermal activation model for $1/f^{\gamma}$ noise in Si-MOSFETs, *Solid-State Electron.* **31**, 959-964 (1988).

40. M. J. Kirton and M. J. Uren, Noise in solid-state microstructures: a new perspective on individual defects, interface states and low-frequency (1/f) noise, *Advances in Physics* **38**, 367-468 (1989).
41. N. V. Amarasinghe, Z. Çelik-Butler, and A. Keshavarz, Extraction of oxide trap properties using temperature dependence of random telegraph signals in submicron metal-oxide-semiconductor field-effect transistors, *J. Appl. Phys.* **89**, 5526-5532 (2001).
42. E. Simoen, A. Mercha, L. Pantisano, C. Claeys, and E. Young, Tunneling $1/f^{\gamma}$ noise in 5 nm HfO_2/2.1 nm SiO_2 gate stack n-MOSFETs, *Solid-State Electron.* **49**, 702-707 (2005).
43. S. T. Martin, G. P. Li, E. Worley, and J. White, The gate bias and geometry dependence of random telegraph signal amplitudes, *IEEE Electron Device Lett.* **18**, 444-446 (1997).
44. J. Koga, S. Takagi, and A. Toriumi, A comprehensive study of MOSFET electron mobility in both weak and strong inversion regimes, in *IEDM Tech. Dig.*, 1994. pp. 475-478.
45. M. Marin, M. J. Deen, M. de Murcia, P. Llinares, and J. C. Vildeuil, Effects of body biasing on the low frequency noise of MOSFETs from 130 nm CMOS technology, *IEE Proc.-Circuits Devices Syst.* **151**, 95-101 (2004).
46. M. Valenza, A. Hoffmann, D. Sodini, A. Laigle, F. Martinez, and D. Rigaud, Overview of the impact of downscaling technology on $1/f$ noise in p-MOSFETs to 90nm, *IEE Proc.-Circuits Devices Syst.*, vol. 151, pp. 102-110, 2004.
47. A. K. M. Ahsan and D. K. M. Schroder, Impact of post-oxidation annealing on low-frequency noise, threshold voltage, and subthreshold swing of p-channel MOSFETs, *IEEE Electron Device Lett.* **25**, 211-213 (2004).
48. J.-S. Lee, D. Ha, Y.-K. Choi, T.-J. King, and J. Bokor, Low-frequency noise characteristics of ultrathin body p-MOSFETs with molybdenum gate, *IEEE Electron Device Lett.* **24**, 31-33 (2003).
49. M. Fadlallah, G. Ghibaudo, J. Jomaah, and G. Guégan, Static and low frequency noise characterization in surface- and buried-mode 0.1 μm PMOSFETS, *Solid-State Electron.* **47**, 1155-1160 (2003).
50. E. Emrani, F. Balestra, and G. Ghibaudo, On the understanding of electron and hole mobility models from room to liquid helium temperatures, *Solid-State Electron.* **37**, 1723-1730 (1994).
51. S. C. Sun and J. D. Plummer, Electron mobility in inversion and accumulation layers on thermally oxidized silicon surfaces, *IEEE Trans Electron Devices* **ED-27**, 1497-1508 (1980).
52. A. Pacelli, S. Villa, A. L. Lacaita, and L. M. Perron, Quantum effects on the extraction of MOS oxide traps by $1/f$ noise measurements, *IEEE Trans. Electron Devices*, **46**, 1029-1035 (1999).
53. S. Dimitrijev and N. Stojadinovic, Analysis of CMOS transistor instabilities, *Solid-State Electron.* **30**, 991-1003 (1987).
54. M. N. Darwish, J. L. Lentz, M. R. Pinto, P. M. Zeitzoff, T. J. Krutsick, and H. H. Vuong, An improved Electron and Hole Mobility Model for General Purpose Device Simulation, *IEEE Trans. Electron Devices* **44**, 1529-1538 (1997).
55. M. von Haartman, J. Westlinder, D. Wu, B. G. Malm, P.-E. Hellström, J. Olsson and M. Östling, Low-frequency noise and Coulomb scattering in $Si_{0.8}Ge_{0.2}$ surface channel pMOSFETs with ALD Al_2O_3 gate dielectrics, *Solid-State Electronics* **49**, 907-914 (2005).
56. W. M. Soppa and H.-G. Wagemann, Investigation and modeling of the surface mobility of MOSFET's from -25 to 150°C, *IEEE Trans. Electron Devices* **35**, 970-977 (1988).
57. J. H. Scofield, N. Borland, and D. M. Fleetwood, Reconciliation of different gate-voltage dependencies of $1/f$ noise in n-MOS and p-MOS transistors, *IEEE Trans. Electron Devices* **41**, 1946-1952 (1994).

58. G. Groeseneken, H. E. Maes, N. Beltrán, and R. F. De Keersmaecker, A reliable approach to charge-pumping measurements in MOS transistors, *IEEE Trans. Electron Devices* **ED-31**, 42-53 (1984).
59. T. Sakurai and T. Sugano, Theory of continuously distributed trap states at Si-SiO$_2$ interfaces, *J. Appl. Phys.* **52**, 2889-2896 (1981).
60. M. S. Kim, I. C. Nam, H. T. Kim, H. T. Shin, T. E. Kim, H. S. Park, K. S. Kim, K. H. Kim, J. B. Choi, K. S. Min, D. J. Kim, D. W. Wang, and D. M. Min, Optical subthreshold current method for extracting the interface states in MOS systems, *IEEE Electron Device Lett.* **25**, 101-103 (2004).
61. M. J. Knitel, P. H. Woerlee, A. J. Scholten, and A. T. A. Zegers-Van Duijnhoven, Impact of process scaling on 1/f noise in advanced CMOS technologies, in *IEDM Tech. Dig.*, 2000, pp. 463-466.
62. R. Brederlow, W. Weber, D. Schmitt-Landsiedel, and R. Thewes, Fluctuations of the low frequency noise of MOS transistors and their modeling in analog and RF-circuits, in *IEDM Tech. Dig.*, 1999, pp. 159-162.
63. A. P. van der Wel, E. A. M. Klumperink, S. L. J. Gierkink, R. F. Waassenaar, and H. Wallinga, MOSFET 1/f noise measurements under switched bias conditions, *IEEE Electron Device Lett.* **21**, 43-46 (2000).
64. R. Brederlow, J. Koh, G. I. Wirth, R. da Silva, M. Tiebout, and R. Thewes, Low frequency noise considerations for CMOS analog circuit design, in *Proc. Int. Conf. Noise and Fluctuations (ICNF)*, 2005, pp. 703-708.
65. P. Srinivasan, E. Simoen, L. Pantisano, C. Claeys, and D. Misra, Impact of gate material on low-frequency noise of nMOSFETs with 1.5 nm SiON gate dielectric: testing the limits of the number fluctuation theory, in *Proc. Int. Conf. Noise and Fluctuations (ICNF)*, 2005, pp. 231-234.
66. T. Contaret, K. Romanjek, T. Boutchacha, G. Ghibaudo, and F. Bœuf, Low frequency noise characterization and modelling in ultrathin oxide MOSFETs, *Solid-State Electron.* **50**, 63-68 (2006).
67. X. Li, C. Barros, E. P. Vandamme, and L. K. J. Vandamme, Parameter extraction and 1/f noise in a surface and a bulk-type, p-channel LDD MOSFET, *Solid-State Electron.* **37**, 1853-1862 (1994).
68. A. van der Ziel, Flicker noise in semiconductors: not a true bulk effect, *Appl. Phys. Lett.* **33**, 883-884 (1978).
69. A. van der Ziel, Unified presentation of 1/f noise in electronic devices: fundamental 1/f noise sources, *Proc. IEEE* **76**, 233-258 (1988).
70. J. Rhayem, D. Rigaud, A. Eya'a, and M. Valenza, 1/f noise in metal-oxide-semiconductor transistors biased in weak inversion, *J. Appl. Phys.* **89**, 4192-4194 (2001).
71. L. K. J. Vandamme and G. Trefán, 1/f noise in homogeneous and inhomogeneous media, *IEE Proc.-Circuits Devices Syst.* **149**, 3-12 (2002).
72. F. N. Hooge and L. K. J. Vandamme, Lattice scattering causes 1/f noise, *Phys. Lett. A* **66**, 315-316 (1978).
73. M. B. Weissman, 1/f noise and other slow, nonexponential kinetics in condensed matter, *Rev. Mod. Phys.* **60**, 537-571 (1988).
74. E. P. Vandamme and L. K. J. Vandamme, Unsolved problems on 1/f noise in MOSFETs and possible solutions, in *Proc. Unsolved Problems of Noise and fluctuations (UPoN)*, 1999, pp. 395-400.
75. L. K. J. Vandamme and S. Oosterhoff, Annealing of ion-implanted resistors reduces the 1/f noise, *J. Appl. Phys.* **59**, 3169-3174 (1986).
76. L. K. J. Vandamme, Noise as a diagnostic tool for quality and reliability of electronic devices, *IEEE Trans. Electron Devices* **41**, 2176-2187 (1994).

77. P. Gaubert, A. Teramoto, T. Hamada, M. Yamamoto, K. Kotani, and T. Ohmi, $1/f$ noise suppression of pMOSFETs fabricated on Si(110) and Si(100) using an alkali-free cleaning process, *IEEE Trans. Electron Devices* **53**, 851-856, (2006).
78. B. Cretu, M. Fadlallah, G. Ghibaudo, J. Jomaah, F. Balestra, and G. Guégan, Thorough characterization of deep-submicron surface and buried channel pMOSFETs, *Solid-State Electron.* **46**, 971-975 (2002).
79. M. Fadlallah, G. Ghibaudo, J. Jomaah, M. Zoaeter, and G. Guégan, Static and low frequency noise characterization of surface- and buried-mode 0.1 μm P and NMOSFETS, *Microelectron. Reliab.* **42**, 41-46 (2002).
80. R. A. Wilcox, J. Chang, and C. R. Viswanathan, Low-temperature characterization of buried-channel NMOST, *IEEE Trans. Electron Devices* **36**, 1440-1447 (1994).
81. S. Okhonin, M. A. Py, B. Georgescu, H. Fischer, and L. Risch, DC and low-frequency Noise Characteristics of SiGe P-Channel FET's Designed for 0.13-mm Technology, *IEEE Trans. Electron Devices* **46**, 1514-1517 (1999).
82. K. Akarvardar, B. M. Dufrene, S. Cristoloveanu, P. Gentil, B. J. Blalock, and M. M. Mojarradi, Low-frequency noise in SOI four-gate transistors, *IEEE Trans. Electron Devices* **53**, 829-835 (2006).
83. M. de Murcia, M. Marin, Y. Akue Allogo, D. Rigaud, P. Llinares, and D. Cottin, Impact of gate engineering and silicidation on low frequency noise characteristics in 0.18 μm technology MOSFETs, in *Proc. Int. Conf. Noise in Physical Systems and 1/f Fluctuations (ICNF)*, 2001, pp. 149-152.
84. M. Marin, J. C. Vildeuil, B. Tavel, B. Duriez, F. Arnaud, P. Stolk, and M. Woo, Can $1/f$ noise in MOSFETs be reduced by gate oxide and channel optimization?, in *Proc. Int. Conf. Noise and Fluctuations (ICNF)*, 2005, pp. 195-198.
85. S. V. Melkonyan, V. M. Aroutiounian, F. V. Gasparyan, and H. V. Asriyan, Phonon mechanism of mobility fluctuation equilibrium fluctuation and properties of $1/f$ noise, *Physica B* **382**, 65-70, (2006).
86. N. W. Ashcroft and N. D. Mermin, *Solid State Physics* (Brooks/Cole Thomson Learning, The United States, 1976).
87. N. Takabatake, T. Kobayashi, Y. Show, and T. Izumi, Photoacoustic evaluation of defects and thermal conductivity in the surface layer of ion implanted semiconductors, *Mat. Sci. Eng.* **B91-92**, 186-188 (2002).
88. T.-L. Hsu, D. D.-L. Tang, and J. Gong, Low-frequency noise properties of dynamic threshold (DT) MOSFET's, *IEEE Electron Device Lett.* **20**, 532-534 (1999).
89. S. Haendler, J. Jomaah, G. Ghibaudo, and F. Balestra, Improved analysis of low frequency noise in dynamic threshold MOS/SOI transistors, *Microelectron. Reliab.* **41**, 855-860 (2001).
90. A. Asai, J. Sato-Iwanaga, A. Inoue, Y. Hara, Y. Kanzawa, H. Sorada, T. Kawashima, T. Ohnishi, T. Takagi, and M. Kubo, Low-frequency noise characteristics in SiGe channel heterostructure dynamic threshold pMOSFETs (HDTMOS), in *IEDM Tech. Dig.*, 2002, pp. 35-38.
91. A. K. M. Ahsan and D. K. Schroder, Impact of channel carrier displacement and barrier height lowering on the low-frequency noise characteristics of surface-channel n-MOSFETs, *Solid-State Electron.* **49**, 654-662 (2005).
92. M. von Haartman, Ph. D. Thesis, KTH, Royal Institute of Technology, Sweden, 2006.
93. A. Mercha, E. Simoen, and C. Claeys, Impact of high vertical electric field on low-frequency noise in thin-gate oxide MOSFETs, *IEEE Trans. Electron Devices*, **50**, 2520-2527 (2003).
94. W. Liu, X. Jin, X. Xi, J. Chen, M.-C. Jeng, Z. Liu, Y. Cheng, K. Chen, M. Chan, K. Hui, J. Huang, R. Tu, P. K. Ko, and C. Hu, BSIM3v3.3 MOSFET model Users' manual, Department of electrical engineering and computer sciences, University of California, Berkeley, CA 94720.

PROBLEMS

1. Derive an expression for mobility fluctuation noise in the saturation region on the form

$$S_{I_D} = A \cdot \frac{I_D^2}{V_{GS} - V_T}$$

where A is a constant. Compare with the expression in the linear region (Eq. 3-80).

2. Derive an expression for the drain current noise PSD as a function of drain current for the number fluctuation noise model in the saturation region. Start from Eq. (3-79) and compare your final result with Eqs. (3-67) and (3-68).

3. Identify the dominant noise sources from the following observed relations between the output drain current noise and the drain current

$$S_{I_D} \propto I_D^k.$$

I_D is varied by the gate voltage while V_{DS} is held constant. k is a current exponent equal to

(a) $k = 1$
(b) $k = 3/2$
(c) $k = 0$
(d) $k = 4$
(e) $k = 2$

Assume for simplicity that $R_S = 0$ and $\alpha = 0$. The noise could either be generated in the channel or in the series resistance R_D. There could also be several alternatives for each k value.

Chapter 4

1/f NOISE PERFORMANCE OF ADVANCED CMOS DEVICES

1. INTRODUCTION

For many decades, the performance of MOS transistors was enhanced mainly by decreasing the gate length of the devices. The gate length has been reduced from hundreds of micrometer in the 1960s to a few tens of nanometer today (2006). Already in the beginning of the 1970s, it was discovered that devices with short gate lengths (around 1 µm at the time) showed an undesired behaviour called the *short-channel effect*.[1] The gate loses control over the channel as the gate length is scaled down, which leads to a reduced threshold voltage and an increased off-current for devices with short channel lengths. The problem was solved by invoking scaling rules stating that several critical transistor dimensions and parameters must be reduced with a common multiplicative factor.[2] However, about a decade ago other techniques to improve device performance became attractive research subjects at the same time as the traditional downscaling again faced difficulties. Moreover, it was found that continued downscaling necessitates use of new dielectric materials for the insulator layer (gate oxide) between the gate and the semiconductor substrate. All of this has resulted in the development and emergence of a broad range of CMOS technologies including new materials and architectures. This shift in technology has several important implications on the low-frequency (LF) noise performance. The LF noise is, as discussed in the previous chapter, sensitive to defects and imperfections in the current path. The device quality in terms of defect densities as well as the LF noise properties may differ substantially for different materials, manufacturing technologies and device architectures.

At the same time as the more complex technology may lead to increased LF noise, the downsizing of the transistor dimensions certainly causes higher LF noise. This is immediately found from Eqs. (3-30), (3-38) and (3-49) and is explained by the fact that the noise variance increases when the number of fluctuating carriers is fewer. The aforementioned discussion highlights the importance of LF noise studies in advanced CMOS transistors and systematic evaluations of the impact of different technologies on the LF noise performance.

This chapter will in detail describe various CMOS technologies that are important today and present the LF noise characteristics for each technology (sections 3 to 9). First, we will further motivate the demand for new materials and device concepts and give an overview of where CMOS stands today in the coming two subsections. Then, the $1/f$ noise performance in CMOS devices is summarized in section 2, preceding the more detailed discussion in sections 3 to 9.

1.1 Demand for new materials and MOSFET concepts

The speed of an electric circuit, for example the maximum clock frequency of a microprocessor, is ultimately limited by the delay of its building blocks, the transistors. The maximum clock frequency of Intel's microprocessors has been boosted from 100 kHz in the first 4004 processor in 1971 to around 4 GHz today, merited by the increase of transistor speed. The main driver of the enhanced performance of CMOS devices has, up to now, been the downscaling of device dimensions. Tremendous advances in fabrication technology, especially lithography techniques, have made the rapid downscaling possible. Fig. 4-1 shows how the minimum MOS transistor feature size in production has evolved during the past three decades. A commonly used figure of merit for the internal delay of a MOS transistor is the CMOS inverter delay given below

$$\tau = \frac{C_G V_{DD}}{I_{D,sat}} \qquad (4\text{-}1)$$

where C_G is the gate capacitance (not per unit area) and V_{DD} is the circuit supply voltage. Inserting Eq. (3-8) for $I_{D,sat}$ and assuming $V_{DD} \gg V_T$ gives $\tau \propto L^2/(\mu_{eff} V_{DD})$. As can be seen, the speed (τ^{-1}) increases quadratically with decreasing gate length. For analog applications, which this book mainly deals with, the transition frequency f_T is of utmost importance. The transition frequency is defined as the frequency where the current gain of the transistor has dropped to unity. Thus,

$$f_T = \frac{g_m}{2\pi C_{GS}} = \{\text{using Eq. (3-8)}\} = \frac{3\mu_{\textit{eff}}(V_{GS}-V_T)}{4\pi m L^2}. \qquad (4\text{-}2)$$

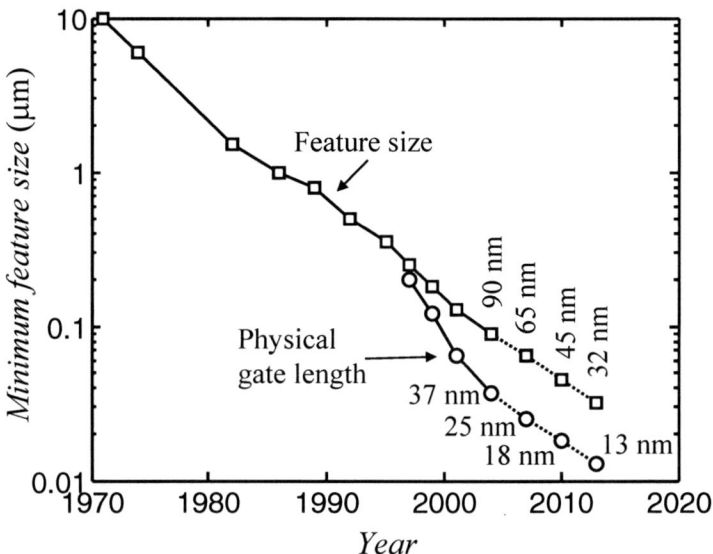

Figure 4-1. MOSFET minimum feature size evolution over time.

Obviously, improved mobility and decreased gate length yield enhanced device speed. However, several unwanted effects degrading the device operation and performance come into play when the device dimensions are reduced. The short-channel effect (SCE) became a problem already in the early 1970s when the shortest gate length was around 1 µm. The gate loses control over the channel as the gate length is scaled down, which leads to a reduced threshold voltage and increased off-current for short channel lengths. Another related effect is the *Drain-Induced-Barrier-Lowering* (*DIBL*), causing a further reduction in the threshold voltage and increase of the off-current when a high drain voltage is applied. One can illustrate this as the source and drain depletion regions take up a larger share of the total depletion region under the channel, less charge is controlled by the gate (charge-sharing model).[3] Three simple remedies to control the SCE and the DIBL are consequently decreased S/D junction depth, increased channel doping and decreased oxide thickness. Decreasing the supply voltage will also alleviate the problem with DIBL. To keep control of the SCE and DIBL, rules involving scaling of several device and circuit parameters with a common multiplicative factor were proposed.[2] The device dimensions t_{ox}, L,

W, x_j are scaled with a factor $1/\kappa$, where $\kappa > 1$. The channel doping concentration is scaled with κ, and the supply voltages with $1/\kappa$. Scaling according to these rules will keep the electric field constant (constant-field scaling). Later a generalized scaling scheme intended to preserve the shape of the electric field was introduced, involving two scaling factors.[4]

Unfortunately, several undesired effects arise from the scaling. The mobility is degraded when the doping concentration is increased due to higher effective electric field as well as due to impurity scattering. Moreover, the source and drain resistance must be scaled down in proportion to the channel resistance, which is increasingly difficult as the junction depth decreases. Reliability and power dissipation become serious problems in the generalized scaling scheme, associated with the increase of the field intensity resulting from this scheme. The tunneling current through the gate oxide increases exponentially with decreasing thickness. Eventually, for physical oxide thicknesses around 1-1.5 nm the gate leakage current becomes unacceptably high.[5] In addition, the depletion of the gate poly-Si (called poly-depletion) gives rise to a reduction of the *effective* oxide capacitance, since the depletion zone appears as a capacitance acting in series with the gate oxide capacitance. All these effects together demonstrate the need to find alternative materials and device architectures to enhance mobility, control the short channel effects, limit the gate leakage current etc. In the remaining part of this chapter, the most important *non-classical CMOS concepts* of today will be presented and analyzed in terms of LF noise performance.

1.2 Advanced CMOS technology overview

The various technological requirements on future semiconductor devices are assessed by the International Technology Roadmap for Semiconductors (ITRS).[5] The scaling requirements are less stringent on the devices for analog/RF applications compared to the case for digital CMOS. On the other hand, additional requirement such as low $1/f$ noise are imposed on the analog/RF devices. The scaling predictions and various technology requirements, including the $1/f$ noise performance, are summarized according the ITRS roadmap for RF and analog mixed-signal CMOS technology in Table 4-1. The shaded areas point out problems for which manufacturable solutions do not exist today. As seen, the $1/f$ noise is predicted to become a big concern from 2010 and on.

Today, CMOS technology dominates in consumer electronics products and can be used for applications below 10 GHz. For higher frequencies, SiGe heterojunction bipolar transistors (HBTs) and transistors based on III-V materials are used. High electron mobility transistors (HEMTs) and HBTs

4. 1/f noise performance of advanced CMOS devices

made in GaAs and InP are the fastest devices today and can be used in extremely demanding applications with operating frequencies around 100 GHz. MOS transistors generally show higher $1/f$ noise than bipolar transistors, and are therefore usually less preferred in low-noise applications. CMOS technology, on the other hand, is superior in terms of low cost, scalability, low standby power, higher integration capability and integrated functions. Due to these reasons and the rapid development of CMOS technology, it is expected that CMOS will gain in importance in the RF/analog domain, especially when the systems require a higher degree of functionality.[5,6]

Table 4-1. RF and analog/mixed-signal CMOS technology requirements according to the ITRS roadmap, 2005 edition (see http://www.itrs.net).

Year of production	2005	2007	2010	2013	2016	2020
DRAM ½ pitch[a] (nm) (Technology node)	80	65	45	32	22	14
Supply voltage[b] (V)	1.2	1.2	1.1	1.0	1.0	1.0
Gate length[c] (nm)	75	53	32	22	16	11
t_{EOT}[d] (nm)	2.2	2.0	1.5	1.3	1.1	0.9
$1/f$ noise[e] ($\mu V^2 \cdot \mu m^2/Hz$)	190	160	90	70	50	30
g_m/g_{ds} at $5 \cdot L_{min\ digital}$[f]	47	32	30	30	30	30
σV_T matching[g] (mV·µm)	6	6	5	5	4	3
Peak f_T[h] (GHz)	120	170	280	400	550	790
Peak f_{max}[i] (GHz)	200	270	420	590	790	1110
NF_{min}[j] (dB)	0.33	0.25	<0.2	<0.2	<0.2	<0.2

(a) Half the distance between cells in a dynamic RAM memory chip.
(b) Nominal supply voltage V_{DD}.
(c) Minimum nominal gate length for low-standby power digital roadmap.
(d) Equivalent oxide thickness (EOT), SiO_2 equivalent physical thickness.
(e) Input gate voltage noise PSD at 1 Hz normalized with gate area, $W \cdot L \cdot S_{Vg}$, measured at $V_{GS} - V_T = 0.1$ V.
(f) Measure of amplification (at low-frequencies) of a 5×minimum gate length low-standby power CMOS transistor. $V_{DS} = V_{DD}/2$ and $V_{GS} - V_T = 0.2$ V.
(g) Threshold voltage matching for near neighbouring devices.
(h) Peak transition frequency.
(i) Peak maximum oscillation frequency.
(j) Minimum transistor noise figure at 5 GHz.
Grey shaded areas point out problems for which manufacturable solutions not exist.

A multitude of different approaches to enhance the performance of CMOS devices (in order for them to meet the requirements) have been suggested and researched. The most important groups are related to the enhancement of mobility (strained Si, SiGe etc) or improved scalability by using silicon-on-insulator (SOI) substrates or multiple gates etc. Fig. 4-2 presents an overview and future directions of CMOS technology.

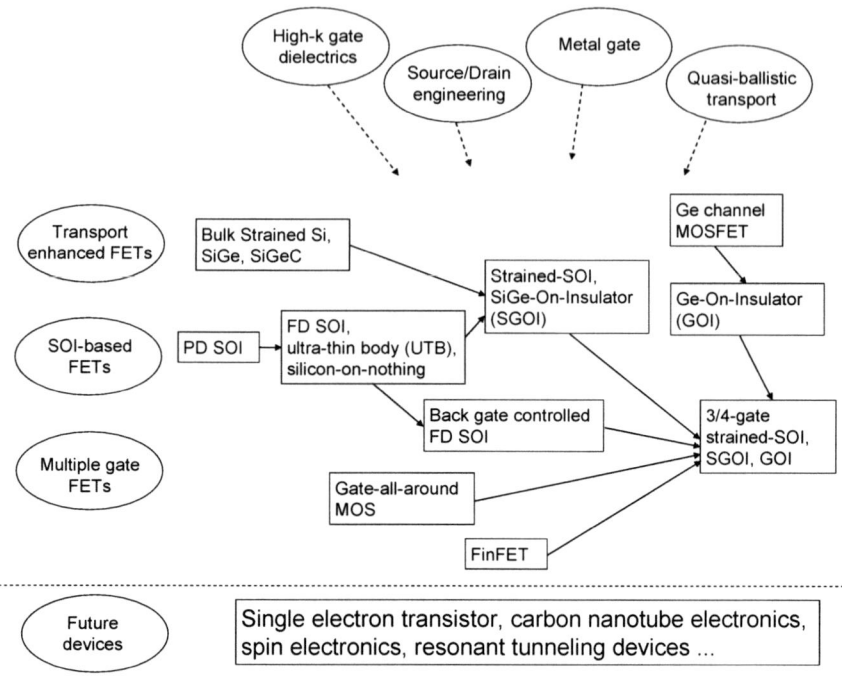

Figure 4-2. CMOS technology overview.

High-k gate dielectrics, metal gate (instead of poly-silicon), source/drain engineering (for example Schottky-junctions in S/D) and quasi-ballistic transport (basically shorter devices) can and must be integrated in the different kinds of FETs for improved performance. This book deals with all of these technologies and different kinds of FETs. The $1/f$ noise is primarily generated in the channel why source/drain engineering may be of minor importance for the $1/f$ noise properties. We will however discuss Schottky Barrier (SB) MOSFETs in the SOI section and show that a careless implementation of Schottky junctions may increase the $1/f$ noise by orders of magnitude. The future devices, on the other hand, are beyond the scope of this presentation. The book edited by A. A. Balandin is recommended to any reader interested in such devices.[7] For example M. N. Mihaila discusses single electron transistors and carbon nanotubes in his interesting chapter about LF noise in nanomaterials and nanostructures.

2. 1/f NOISE PERFORMANCE OVERVIEW

The downscaling of the supply voltage and the device dimensions cause a degradation of the signal-to-noise ratio. Moreover, the shift to more complex device structures and/or new materials has so far resulted in elevated LF noise levels. It has even been suggested that the LF noise, due to its inverse dependence on gate area, can be a showstopper for CMOS scaling in certain applications.[8] The main topic of this chapter is to investigate the LF noise properties of advanced CMOS devices that may reside in future analog and digital applications and improve the understanding of the physical mechanisms causing the LF noise. The collected experiences can serve as a guideline how to design a device for low noise and which materials and structures that should be avoided. Further downscaling of device dimensions will increase the $1/f$ noise, which makes it extremely important to understand the origin of the noise and reduce it by clever design.

The $1/f$ noise can be strongly dependent on the CMOS technology since defects and imperfections in the current path and close to it, especially traps in the gate oxide, have a detrimental impact on $1/f$ noise performance. Technology considerations such as choice of gate oxide and channel material as well as the fabrication process in terms of cleaning, oxidation and deposition are therefore of high importance. The device architecture is also of vital importance. It is established that a buried channel for the current transport is beneficial for low $1/f$ noise. The trap density N_t and Hooge parameter α_H that were defined in the previous chapter can both be used as figures of merit for the $1/f$ noise performance irrespective of the origin of the noise. Fig. 4-3 summarizes extracted and published values (or calculated from published data) of (a) N_t and (b) α_H for various CMOS technologies.[9-46]

The future requirements on N_t and α_H have been calculated from the ITRS roadmap by using the input gate voltage noise values given in Table 4-1 and are plotted as broken lines in Fig. 4-3. As seen, the high-k gate dielectric devices as well as the SOI MOSFETs have problems with meeting the requirements, hence technology improvements are necessary. The dispersion in the data spans three decades for the conventional CMOS devices. The Si n- and pMOSFETs showing the lowest N_t and α_H values use pure SiO_2 as gate oxide, some of them have a buried channel (surface of the substrate counterdoped), while the highest values are for devices with a nitrided gate oxide. Note that N_t is more commonly reported than α_H, especially for nMOS and SOI transistors.

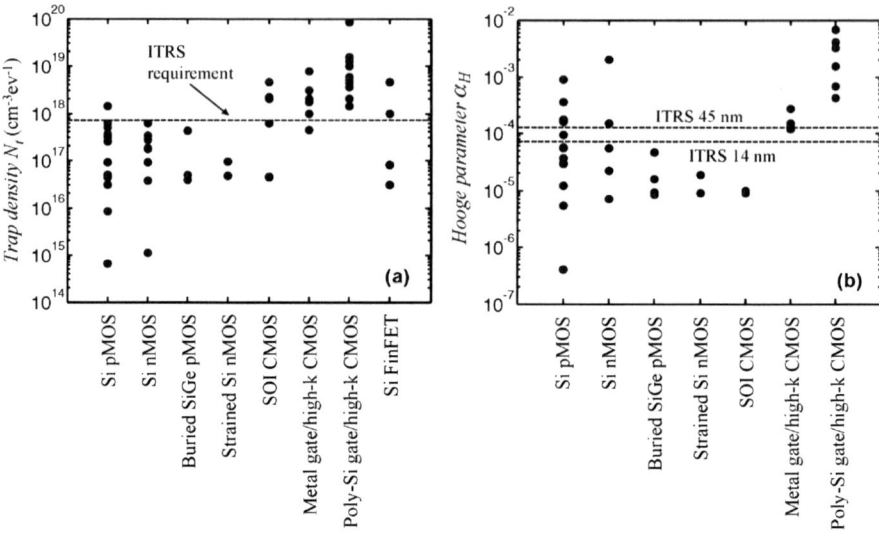

Figure 4-3. Summary of oxide trap density (N_t) and Hooge parameter (α_H) values reported in literature.[9-46]

3. 1/f NOISE PERFORMANCE OF ULTRA SCALED DEVICES

3.1 Consequences of small geometrical dimension

The normalized drain current noise PSD varies inversely with gate area for both number and mobility fluctuation noise according to Eqs. (3-30), (3-38) and (3-49). Further downscaling of the gate length therefore leads to increased $1/f$ noise. However, the oxide thickness is also downscaled, which either can worsen or improve the $1/f$ noise performance. For the number fluctuation noise model

$$S_{V_G} \propto 1/(W \cdot L \cdot C_{ox}^2) \propto t_{EOT}^2/(W \cdot L) \qquad (4\text{-}3)$$

while

$$S_{V_G} \propto 1/(W \cdot L \cdot C_{ox}) \propto t_{EOT}/(W \cdot L) \qquad (4\text{-}4)$$

in Hooge's model. Thus, lower noise is expected for MOSFETs with thinner oxides according to the models. Fig. 4-4 shows a simulation of the

4. 1/f noise performance of advanced CMOS devices

1/f noise scaling using the predicted geometrical values in Table 4-1. Unfortunately, the trap density and Hooge parameters cannot be considered as constants, they usually increase for smaller t_{EOT}, due to technological reasons. Nitridation of the SiO_2 gate oxide forming oxynitride (SiON) is necessary in order to scale down the oxide thickness for the technology nodes beyond 0.18 μm ($t_{ox} \sim 3.5$ nm). The nitrogen incorporation into the SiO_2 introduces charged traps in the oxide, which cause higher 1/f noise. Further on, beyond the 65 nm node, the nitrided SiO_2 is no longer sufficient and must be replaced with high-k gate dielectrics in order to control the gate leakage current. The high-k gate dielectric MOSFETs show even worse noise performance, which will be addressed later in this chapter.

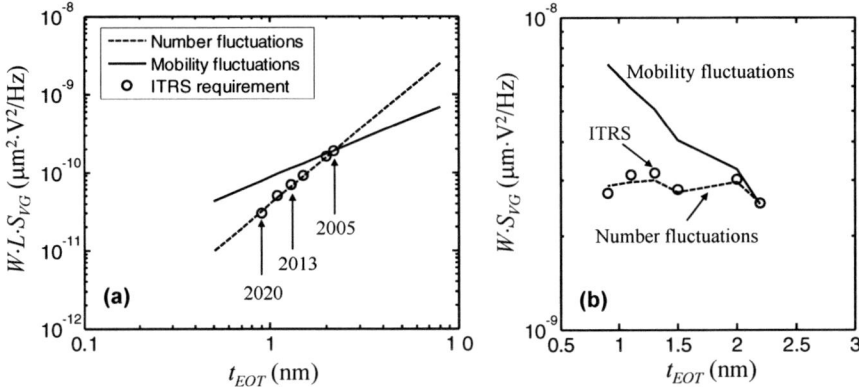

Figure 4-4. Predicted evolution of the input gate voltage noise with equivalent oxide thickness; (a) shows the input gate voltage noise normalized to gate area and (b) normalized to gate width. The 1/f noise value given for year 2005 in the ITRS roadmap (see Table 4-1) was taken as a reference value for the calculations.

An elevation of the average noise level is not the only concern for downscaled devices. The device-to-device variations in the 1/f noise performance increase when the gate area is reduced. Statistical fluctuations in the number of traps among an ensemble of devices will have a large impact on the 1/f noise level when the devices are so small that only a few traps are present.[47,48] The relative standard deviation of the number of traps is given by Brederlow et al. as[48]

$$\sigma_{N_t} = 1/\sqrt{N_t \cdot W \cdot L}. \tag{4-5}$$

Even if the average 1/f noise is below the required limit, some devices could display much higher noise. The modeling of the 1/f noise becomes more difficult; one has to include also the standard deviation in the models.[49]

An interesting situation occurs if a device contains no traps at all. This could happen for very small devices (gate area below around 0.1 μm^2). The number fluctuation noise is completely eliminated in such case, which obviously can lower the LF noise drastically. However, the mobility fluctuation noise still remains. This is evident from experiments of RTS noise in small devices, where the residual signal after subtraction of the RTS noise still exhibits a $1/f$ spectrum.[50] If one can afford to characterize and select the low-noise devices (optimally without traps) for making circuits, improved noise performance can be achieved. However, a problem arises if a trap suddenly is generated during operation, which directly would compromise the low-noise behaviour. What about the mobility fluctuation noise in ultra short devices? One interesting phenomena in ultra short devices in the 10-nm range and below is the occurrence of ballistic transport; some carriers might traverse from source to drain without being scattered. Since the $1/f$ noise is generated from the involvement of phonons when they scatter with electrons, the $1/f$ noise could possibly decrease in the case of ballistic transport.

Further effects that come into play in aggressively scaled transistors are defects at the edges in narrow transistors leading to higher $1/f$ noise,[51] hot carriers which may cause degradation of the gate oxide and thus increased noise,[52] and field-dependence on the Hooge parameter.[53]

3.2   Nitrided gate oxides

The major advantages with oxynitrides are prevention of boron penetration through the gate oxide and improved hot carrier reliability. The dielectric constant is also slightly higher. The dielectric constant of Si_3N_4 and SiO_2 is 7 and 3.9, respectively. Oxynitrides will have a value in between those, depending on the nitrogen content. However, the mobility reduces due to Coulomb scattering and the $1/f$ noise increases in devices with nitrided oxides. A small mobility reduction at low electric field is usually observed for both electrons and holes due to the increased oxide charge.[54,55] At high electric fields, on the other hand, electrons and holes show different behaviours. It has been reported that the electron surface roughness mobility is increased in oxynitrides, whereas a reduction is found for holes, which was explained by a change in nature of the interface geometry by the NO oxidation.[56]

It is now well established that MOSFETs with nitrided gate oxide exhibit higher $1/f$ noise than that for devices with pure SiO_2. Morfouli et al. reported that N_t was found to increase exponentially with the nitrogen content in the range between 0 and 11%.[57] P-channel MOS transistors are more affected by the nitrogen than the n-channel ones; Da Rold et al. reported a factor of 7-8 increase in $1/f$ noise for pMOS and 2-3 for nMOS with NO annealed oxides.[58] In the same study, a positive charge density equal to 8.7×10^{11} cm^{-2} was found to be introduced by the nitrogen in the pMOSFETs, whereas the nMOSFETs received an extra charge density of 2.9×10^{11} cm^{-2}. The observed correlation between the $1/f$ noise level and the oxide charge density indicated that oxide traps were responsible for the $1/f$ noise. However, some issues are still unexplored. As mentioned in chapter 3.3.3, the type of trap determines if the correlated mobility fluctuations act to increase or decrease the $1/f$ noise. A positive oxide charge indicates a negative correlation between the number and mobility fluctuations for nMOS and a positive correlation for pMOS, but a positive correlation is observed for both device types in the works by Morfouli et al. and Da Rold et al. The oxide/channel interface properties and the mobility are also affected by the NO oxidations, which may result in an increase of the mobility fluctuation noise. This could be interesting topics for future work.

Trap densities above 10^{18} cm^{-3}eV^{-1} are often found in MOSFETs with nitrided gate oxides, which could disenable their use in analog applications. However, the $1/f$ noise can be reduced by locating the nitrogen peak away from the oxide/channel interface, which can be achieved by plasma nitridation[59] or O$_2$ re-oxidation.[58]

3.3 Noise in gate leakage current

Another problem related to the downscaling of the gate oxide thickness is the escalating gate leakage current, which increases exponentially with decreasing t_{ox}. A high gate leakage current is a problem concerning device reliability and also leads to higher power consumption. Moreover, $1/f$ noise and shot noise in the gate leakage current can dominate the output drain current noise under such conditions. Valenza et al. found that the $1/f$ noise in the gate current of a pMOSFET from a 90 nm CMOS technology with t_{ox} = 1.5 nm gave a significant contribution to the drain current noise.[20] Thus, noise stemming from the gate leakage current is becoming a major problem in present and future CMOS technologies. Contaret et al. suggested the following model for the influence of the gate leakage current[60]

$$\frac{S_{I_{D,tot}}}{I_D^2} = \frac{S_{I_D}}{I_D^2} + \kappa_D^2 \times \frac{S_{I_G}}{I_D^2}$$

$S_{I_{D,tot}}$ = total drain current noise PSD (4-6)

S_{I_D} = channel noise PSD

S_{I_G} = gate current noise PSD.

The parameter κ_D is a gate-partitioning coefficient at the drain side ranging between 0 and 1 and is calculated from the current relations below

$$I_{D,ch} = I_{D,meas} + \kappa_D \cdot I_G = I_{S,meas} - \kappa_S \cdot I_G. \qquad (4\text{-}7)$$

The currents $I_{D,ch}$, $I_{D,meas}$, $I_{S,meas}$ and I_G are the intrinsic channel current, measured drain current, measured source current and gate current, respectively. The model in Eqs. (4-6) and (4-7) describes how the noise in the gate leakage current couples out to drain current noise. A physical model of the LF noise in the gate leakage current was developed by Lee and Bosman, the interested reader is referred to literature[61-64] for further details.

4. SiGe CHANNEL pMOSFETs

4.1 Device structure and characteristics

The mobility in the MOSFET channel can be enhanced by channel engineering, which is a common name for a variety of different techniques that aim to improve the channel mobility. The technique of straining the semiconductor lattice is the key concept in realizing enhanced mobility and means that mechanical stress is built in the material. By introducing strain, the energy bands are shifted and distorted which lead to changes in the transport properties. For certain types of strain, if it is tensile or compressive and placed along a certain crystal axis or plane, the net result on the mobility can be positive. One method in order to achieve the strain is to utilize a heterostructure composed by Si and Ge: SiGe. The SiGe can be used as the substrate (see further section 5) to introduce the strain in the Si channel or can be used as the channel itself.

$Si_{1-x}Ge_x$ has a larger lattice constant than Si, varying with Ge composition x between 5.43 Å (Si) to 5.66 Å (Ge). When SiGe is epitaxially grown on Si, the larger lattice constant of the former results in a biaxial compression of the SiGe layer for it to fit with the Si lattice. There is a critical thickness of the layer and a maximum Ge concentration, which depend on each other, for the strain in the SiGe layer to remain.[65] Beyond this limit, undesired relaxation occurs. A schematic structure of a pMOSFET utilizing a buried compressively strained SiGe channel is illustrated in Fig. 4-5(a). A schematic presentation of the lattice distortion and the band diagram of the structure are shown in Figs. 4-5(b) and (c). The bandgap of Ge is smaller than that for Si (0.66 eV compared to 1.12 eV). The bandgap of the compressively strained $Si_{1-x}Ge_x$ is therefore smaller than for Si and varies with x according to[66]

$$E_g(x) = 1.12 - 0.896x + 0.396x^2 \text{ [eV]} \quad (x < 0.3). \tag{4-8}$$

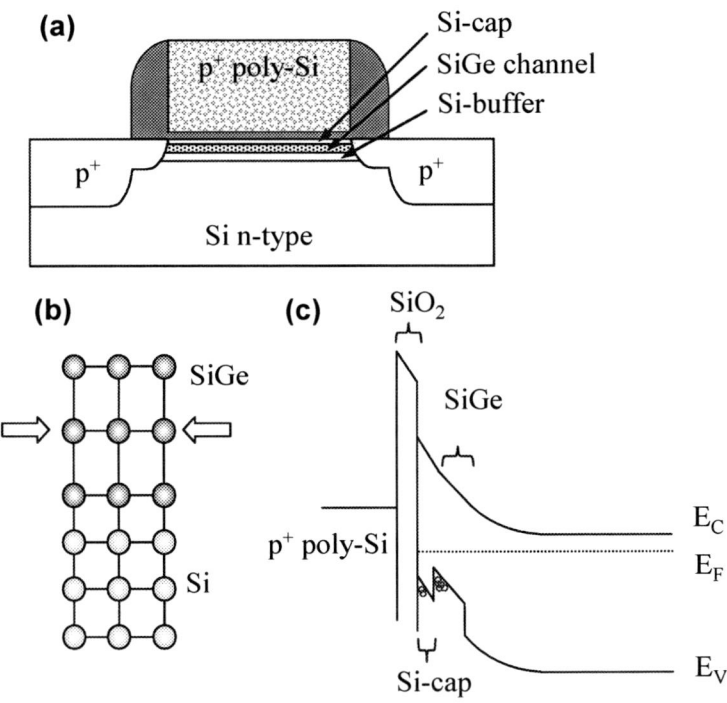

Figure 4-5. (a) Schematic cross section of a buried SiGe channel pMOSFET. (b) Schematic description of the lattice distortion in a SiGe/Si system. (c) Energy band diagram of a MOS structure with a buried SiGe layer.

The compressive strain introduces splits and distortions in the energy band spectrum, which result in lower effective hole mass in the valence band of the SiGe and therefore enhanced hole mobility.[67,68] Fig. 4-6 demonstrates the measured hole mobility of two compressively strained SiGe channel pMOSFETs along with their Si counterparts.

Around 97% of the bandgap offset between Si and SiGe is situated in the valence band.[69] This property can be exploited in a buried SiGe channel pMOSFET, confining holes in a quantum well away from the Si/SiO$_2$ interface, see Fig. 4-5(c). This further enhances the hole mobility due to lower surface scattering. Lower 1/f noise has also been observed for buried SiGe pMOSFETs, which is an additional benefit of this type of device. It should be mentioned that the Si-cap on top of the SiGe channel is necessary also to maintain a low interface state density, since oxidation of SiGe introduce traps that deteriorate the interface.[70,71]

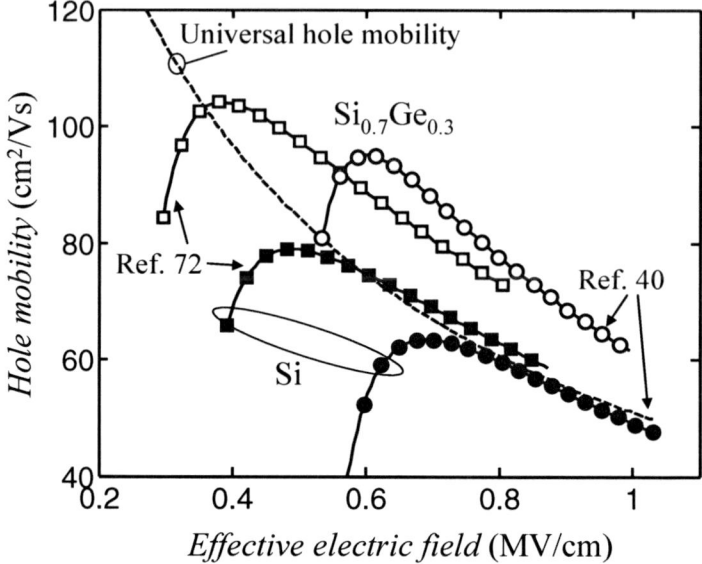

Figure 4-6. Hole mobility vs. effective electric field for two $Si_{0.7}Ge_{0.3}$ pMOSFETs compared with the corresponding reference Si devices.

However, several reports have shown that the hole mobility enhancement in the SiGe channel is reduced when the channel length is decreased, which is a key problem that remains to be solved. Several explanations have been suggested: velocity saturation,[11,73] more pronounced effects of pocket implantations for SiGe,[73-75] longer electrical channel length for SiGe due to reduced boron diffusion,[76] and strain relaxation effects for shorter channels.[74,77] Recently, drive current and transconductance enhancements down to 50-nm gate length have been demonstrated by using SiGe.[72] In the report, it was found that the transconductance increased with decreasing width of the SiGe channel, suggesting strain improvements from the field oxide or local loading effects altering the thickness and/or composition of the Si-cap/SiGe/Si-buffer stack in such a way that it improves the hole confinement in the SiGe channel. Integration of SiGe with SOI technology has also shown very promising results down to 50 nm channel length.[78] The utilization of multiple SiGe well structures,[77] a narrow channel width, or SiGe on SOI are alternative ways forward to achieve enhanced performance also at sub-100 nm gate lengths. Optimizing the Si-cap thickness is also important, both for drive current and noise performance. The Si-cap degrades the gate control of the SiGe channel; the effective gate oxide capacitance is lower as the capacitance of the Si-cap, ε_{si}/t_{cap}, acts in series with the oxide capacitance. Furthermore, a parasitic low-mobility channel is formed in the Si-cap, lowering the overall mobility of the transistor at high gate voltage overdrives.[69] For high drive current the Si-cap should be as thin as possible without causing degradation of the interface by a high density of traps. However, for low $1/f$ noise other considerations prevail, which is discussed in the next subsection.

4.2 $1/f$ noise characteristics

Buried SiGe channel pMOSFETs often exhibit significantly lower $1/f$ noise than that in conventional surface channel Si pMOSFETs. The enhanced mobility, higher intrinsic gain g_m/g_{ds} and the decreased $1/f$ noise make SiGe pMOSFETs attractive for future analog CMOS applications. A field-effect transistor where the current path is separated from the oxide/semiconductor interface is referred to as a buried channel transistor. The $1/f$ noise reduction in buried SiGe channel pMOSFETs is in line with the results on other buried channel devices. Lower $1/f$ noise in comparison with surface channel MOSFETs has also been observed for JFETs and buried Si channel MOSFETs fabricated by counter doping the surface of the substrate (n-type for nMOS and p-type for pMOS), as mentioned in the previous chapter. Several groups have reported lower $1/f$ noise with a factor between 4 and 10 in buried SiGe channel pMOSFETs.[9-11,78-80] However, a reduction is not always observed.[81] In the work performed at KTH from 2002 to 2005, lower $1/f$ noise has also been achieved and promising results presented both for SiGe channel devices on bulk Si as well as on SOI.[40-42] A reduction by a factor of two was found in SiGe devices with a "medium" thick Si-cap. Note in Fig. 4-3 that the $1/f$ noise performance of the best surface Si channel pMOSFETs is at the same level (or lower) as the buried SiGe channel pMOSFETs. One reason behind the significant reduction of the $1/f$ noise for the SiGe devices in comparison with the Si references reported in some work may actually be due to poor noise performance of the references. For Si pMOSFETs with $\alpha_H \sim 10^{-5}$ or $N_t \sim 5\times10^{16}$ cm^{-3}eV^{-1} one can probably expect a less pronounced noise reduction by employing a buried SiGe channel.

Three different origins of the $1/f$ noise reduction in buried SiGe channel pMOSFETs have been suggested. Mathew *et al.* explain the lower noise in the buried channel devices with a lower trap density at the quasi-Fermi level.[10] The valence band offset between Si and SiGe leads to a lower surface potential at the oxide/Si interface than in a surface channel device biased at the same gate voltage overdrive. The interface traps are typically distributed with a higher density close to the valence and conduction band edges and lower density in the middle of the bandgap ("U-shaped"). The Ge-induced valence band offset increases the separation of the quasi-Fermi level from the valence band, which reduces the number of traps that are active in the noise generation if the oxide traps are assumed to be U-shaped in energy. This model has been adopted by other research groups as well, for example Prest and co-workers[11] to mention one. However, it has not been established that the oxide trap density varies appreciably with energy as the model by Mathew et al. requires. One consequence of the model is that decreasing the Si-cap thickness and increasing the Ge-content would result in a stronger noise reduction. However, this is seldom observed for a wide range of Ge concentrations and Si-cap thicknesses. In this work we found the best 1/f noise performance for a Si-cap thickness around 5-6 nm, devices with both thicker and thinner cap showed higher 1/f noise.[40] A similar conclusion was found by Prest et al. Ghibaudo and Chroboczek[81] as well as Tsuchiya et al.[79] report higher 1/f noise for the highest Ge-fractions used in their experiments compared to devices with a moderate Ge-fraction. It should be noted that the density of interface states increases for thin cap layers and high Ge-fractions, which could explain the observed behaviour. However, by biasing the surface Si and buried SiGe channel pMOSFETs in an appropriate way so that the quasi-Fermi levels in the two devices are equally distant from the valence band edges, the extracted trap densities would be roughly equal. In a recent publication,[82] Prest and co-workers were successful with this kind of exercise by including a Hooge mobility fluctuation noise term ($\alpha_H = 2\times10^{-5}$) also, which was not considered in earlier work.

Another model based on correlated mobility fluctuations was presented by Ghibaudo and Chroboczek. They assert that the correlated mobility fluctuations related to Coulomb interaction between the trapped carrier and the channel carriers diminish in the buried channel devices due to the large separation between the traps and the channel. The drain current noise PSD can then be written[81]

$$S_{I_D} = S_{V_{fb}} [g_{m,cap}(1 + \alpha C_{ox}\mu_{eff,cap}I_{D,cap}/g_{m,cap}) + g_{m,SiGe}(1 + R\alpha C_{ox}\mu_{eff,SiGe}I_{D,SiGe}/g_{m,SiGe})]^2 \quad (4\text{-}9)$$

where the subscripts 'cap' and 'SiGe' refer to the cap and the SiGe channel, respectively. The parameter R is a reduction factor around 0.1-0.2. This model can successfully explain a 1/f noise reduction for a buried SiGe channel device biased in strong inversion and is supported in the work by Myronov et al.,[83] but fails below threshold since correlated mobility

4. 1/f noise performance of advanced CMOS devices

fluctuations are unimportant in this region of operation. In fact, the strength of the correlated mobility fluctuations is under debate; see our discussion in chapter 3.3.3. Moreover, Prest et al. could not explain their data using only correlated mobility fluctuations.[82]

A third alternative to explain the lower 1/f noise in buried SiGe pMOSFETs is to invoke mobility fluctuations. The fact that Hooge mobility fluctuation noise often is found to be the dominant 1/f noise source in pMOSFETs was neglected in the previous models. The mobility fluctuation noise is sensitive to the "quality" of the semiconductor material. In chapter 3.4.3.4, we discussed the possibility that (i) random phonon interactions with defects, surfaces and other phonons or (ii) fluctuations in the surface roughness scattering are causing the mobility fluctuation noise. If the carriers are in close proximity with the gate oxide, the 1/f fluctuations in the mobility are expected to be higher than in the bulk. A buried SiGe pMOSFET have two conducting channels; the high-mobility buried SiGe channel and a low-mobility parasitic channel in the Si-cap. A very simple approximation of the carrier distributions in the SiGe and Si channels is

For $V_T \leq V_{GS} \leq V_{T2}$

$$Q_{i,SiGe} = C_{ox,SiGe}(V_{GS} - V_T)$$
$$Q_{i,cap} = 0 \qquad (4\text{-}10)$$

for $V_{GS} \geq V_{T2}$

$$Q_{i,SiGe} = C_{ox,SiGe}(V_{T2} - V_T)$$
$$Q_{i,cap} = C_{ox}(V_{GS} - V_{T2}) \qquad (4\text{-}11)$$

where $1/C_{ox,SiGe} = 1/C_{ox} + t_{cap}/\varepsilon_{si}$.

The Hooge mobility fluctuations can be described by different Hooge parameters for the two channels since uncorrelated noise currents can be assumed[81]

$$\frac{S_{I_D}}{I_D^2} = \frac{q\alpha_{H,cap}}{fWLQ_{i,cap}} \frac{I_{D,cap}^2}{I_D^2} + \frac{q\alpha_{H,SiGe}}{fWLQ_{i,SiGe}} \frac{I_{D,SiGe}^2}{I_D^2}. \qquad (4\text{-}12)$$

The normalized drain current noise is plotted versus gate voltage overdrive V_{GT} for a buried SiGe channel pMOSFET in Fig. 4-7. As seen, the normalized drain current noise flattens out at $V_{GT} \approx 0.2$ V, which is due to the onset of the parasitic channel. The solid line in Fig. 4-7 is a simulation using Eq. (4-12) with $\alpha_{H,cap} = 7\times10^{-5}$, $\alpha_{H,SiGe} = 1.5\times10^{-5}$ and $|V_{T2} - V_T| = 0.25$ V. The $1/f$ noise generated in the parasitic current dominates in this case at a $V_{GT} \geq 0.4$ V. Reducing the Si-cap thickness will postpone the onset of the parasitic current to higher V_{GT}. A thin Si-cap would therefore be desired both to obtain a high drive current and low noise. However, as the SiGe channel is moved closer to the gate oxide interface, the noise generated in the SiGe channel increases.

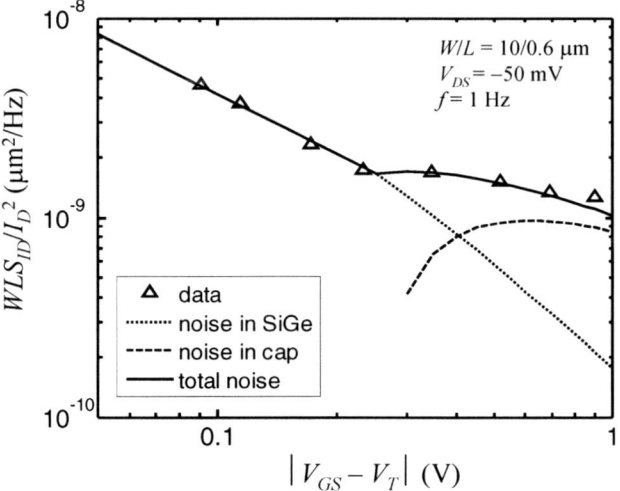

Figure 4-7. Measured (symbols) and simulated (solid or broken lines) drain current noise for a buried $Si_{0.7}Ge_{0.3}$ channel pMOSFET with 5-nm thick Si-cap. Eqs. (4-10) - (4-12) were used in the simulations.

In Fig. 4-8(a), the normalized drain current noise is plotted for SiGe pMOSFETs with different thicknesses of the Si-cap. The normalized drain current noise falls off roughly as $1/V_{GT}$ or weaker, which indicates that mobility fluctuation noise is the origin of the $1/f$ noise. The device with a 3-nm thin Si-cap shows very similar $1/f$ noise level as the Si device indicating that a too thin Si-cap can deteriorate the noise performance of the SiGe channel. The device with a 7-nm thick Si-cap is likely dominated by the $1/f$ noise in the parasitic current at very low $V_{GT} \geq 0.1$ V since the S_I/I^2 curve closely follows the one for Si. A $1/f$ noise reduction with a factor of two was observed for the devices with a medium thick Si-cap (5-6 nm), which optimize the trade-off between sufficient distance from the notorious Si/SiO_2 interface and low parasitic current. It is interesting to note that significantly lower $1/f$ noise was found in a SiGe pMOSFET fabricated on a low-doped

4. 1/f noise performance of advanced CMOS devices

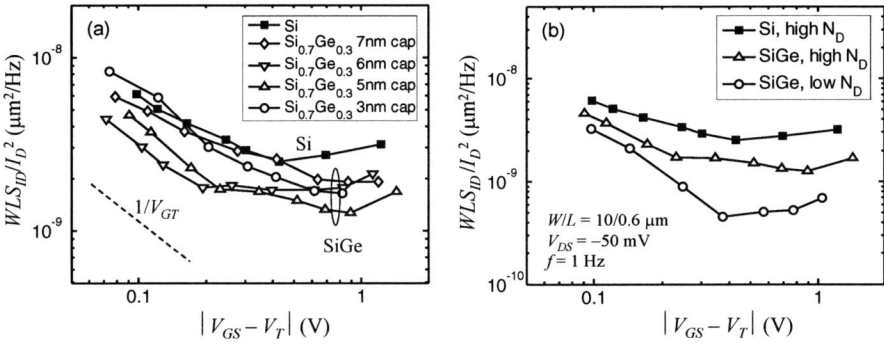

Figure 4-8. Drain current noise measured in SiGe pMOSFETs with (a) different Si-cap thicknesses (reprinted from von Haartman et al.[40] with permission from IEEE), and (b) different substrate doping concentration.

substrate (~10^{14} cm^{-3}) without well or channel implantations as shown in Fig. 4-8(b). The other SiGe devices were implanted with As to a peak concentration of 2×10^{18} cm^{-3}. The lower doping results in a better hole confinement in the SiGe channel since the voltage drop in the Si-cap is reduced. As seen in Fig. 4-8(b), the parasitic channel turns on at a higher V_{GT} than in the other SiGe devices. Obviously, a too low doping leads to problems with short channel effects in standard bulk MOSFETs, but the concept can be employed in SOI MOSFETs.

The density of interface states and the normalized drain current noise extracted at V_{GT} = 0.2 V are plotted versus Si-cap thickness in Fig. 4-9. One possible origin behind the increased 1/f noise for the devices with thin Si-cap could be the degradation of the interface. Direct oxidation of SiGe results in the creation of interface traps.[70-71] Ge atoms segregate and diffuse to the interface during the epitaxy step and the following high-temperature steps, more Ge atoms will reach the interface as the distance is shortened. However, the density of interface states is always lower in the Si device than in the SiGe ones in this case according to Fig. 4-9, which indicates that the interface traps are not likely to be the main origin of the 1/f noise.

Note that the Hooge parameter for the Si pMOSFETs (see Fig. 4-8) varies with gate voltage according to $\alpha_H = 2.9\cdot10^{-5}\cdot(V_{GT}/0.1)^{0.38}$. In other works, such behaviour has often been assumed to be due to a trap density that increases towards the valence band edge, but in our opinion it could be explained by a Hooge parameter that depends on the effective electric field. As explained in chapter 3.5, the position of the inversion carriers depends on the surface electric field perpendicular to the channel direction, which can be varied by the gate and substrate voltages. The results strongly indicate that

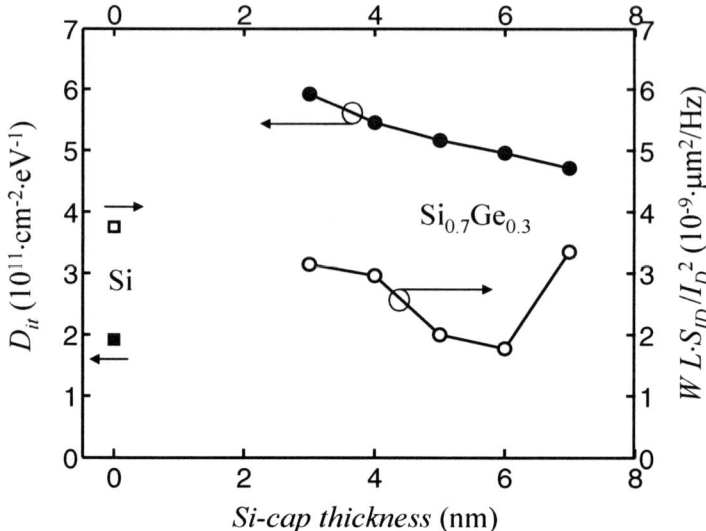

Figure 4-9. Density of interface states and normalized drain current noise (extracted at $V_{GT} = 0.2$ V and $f = 1$ Hz) plotted for SiGe pMOSFETs with different Si-cap thickness. A Si pMOSFET is also included for comparison.

moving the carriers closer to the interface result in an increase of the $1/f$ noise, and vice versa, as was discussed in chapter 3.4.3.3. A consequence of the E-field dependence of the $1/f$ noise is that the Hooge parameter is expected to increase with increasing V_{GT}.

A very high hole mobility can be obtained by forming a high-Ge content channel. A maximum effective hole mobility in the range of 500 cm^2/Vs has been reported by using a $Si_{0.3}Ge_{0.7}$ channel.[83] The carrier confinement improves as the Ge content is increased resulting in a smaller current in the noisy Si-cap. Thus, the $1/f$ noise can be decreased also at high gate voltage overdrives in the SiGe pMOSFETs. A $1/f$ noise reduction by a factor 3 was found by Myronov *et al.* at $V_{GS} - V_T = -1.5$ V, however at the expense of increased $1/f$ noise at lower gate bias.[83] Further mobility improvements can be achieved by utilizing a pure Ge channel. However, initial investigations show that the $1/f$ noise performance of MOSFETs fabricated on Ge-on-insulator substrates is inferior so far.[84]

In summary, buried SiGe channel pMOSFETs can exhibit lower $1/f$ noise than in Si pMOSFETs fabricated with the same technology. A moderate reduction by a factor of two was obtained in our work, which likely can be improved by optimization of the SiGe channel and Si-cap thickness. There are two competing noise mechanisms in a buried SiGe channel pMOSFET:

(i) the 1/ƒ noise generated in the buried channel which decreases as the channel is positioned further from the SiO$_2$/Si interface, and (ii) the 1/ƒ noise generated in the noisy Si-cap which decreases as the confinement in the SiGe channel is improved by, for example, lower doping and/or thinner Si-cap. Buried channel pMOSFETs are advantageous to use in low-power analog applications in view of the low noise at small V_{GT}, high mobility and enhanced intrinsic gain g_m/g_{ds}.

5. STRAINED SI DEVICES

5.1 Strained Si technology and device characteristics

Enhancements of electron and hole mobilities in Si inversion layers from their universal values are possible by strain engineering.[85,86] In contrast, the previous section discussed improved mobility by introducing a strained SiGe channel. Strain enhanced Si-channel mobility in MOSFETs can be realized by local strain techniques such as using silicon nitride capping layer (nMOS) or compressively strained SiGe films in the source/drain regions (pMOS)[86] as well as global techniques such as strained Si on relaxed SiGe virtual substrates[87,88] or strained Si on SiGe-on-insulator.[89] Techniques to obtain improved MOSFET channel mobility is currently a hot research subject; there are several degrees of freedom and options in the type of strain, channel direction and crystal direction to achieve enhanced performance both for nMOS and pMOS devices.[90-92]

In the following, we will describe the properties of MOSFETs with a tensile strained Si-channel. Both the electron and the hole mobility are enhanced by tensile strain induced in the thin epitaxial Si layer grown on a relaxed SiGe virtual substrate as shown in Fig. 4-10. The hole mobility enhancement is often found to drop at high electric fields,[93] but may be maintained at high stress levels (high Ge content) as seen in Fig. 4-10 or by using uniaxial compressive strain instead of biaxial tensile as reported by Thompson et al.[86] The tensile strain causes an energy splitting of the 6-fold degenerate conduction band, resulting in a repopulation of the energy bands that preferentially fills the 2-fold band with lower energy and reduced effective mass. In the valence band, the tensile strain induces energy shifts and warping of the light and heavy hole bands, leading to a reduced effective hole mass at low electric fields. At high stress levels, the energy shifts and the resulting carrier repopulation are correspondingly larger. This leads to reduced interband scattering, which together with the reduced effective masses result in enhanced mobilities.[86,90,94]

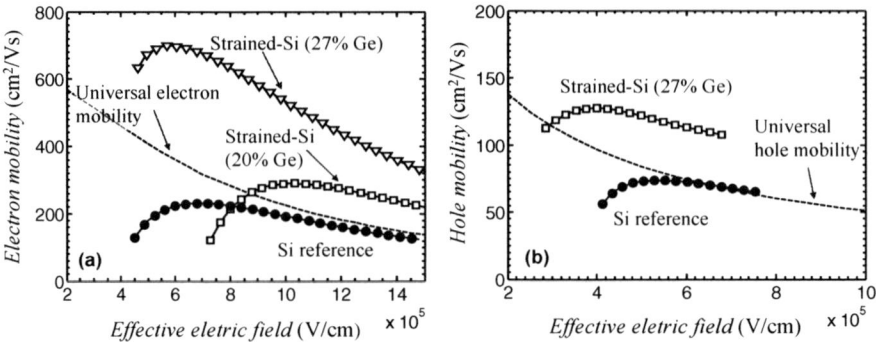

Figure 4-10. (a) Electron and (b) hole mobilities for tensile strained Si MOSFETs fabricated on SiGe virtual substrates with 20% or 27% Ge.

A schematic structure of the strained Si MOSFETs as well as a description of strain effects of the energy bands are presented in Fig. 4-11. The strain-induced electron mobility enhancement is stronger for long channel lengths, as was the case for compressively strained SiGe channel pMOSFETs, but drive current improvements even in sub-50 nm gate length MOSFETs have recently been observed.[95]

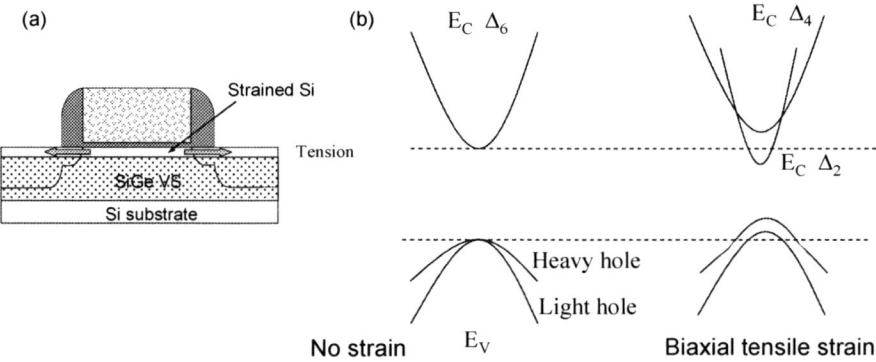

Figure 4-11. (a) Schematic structure of a MOSFET fabricated on a relaxed SiGe virtual substrate. A thin tensile strained Si layer is grown on top of the relaxed SiGe. (b) Schematic description of the energy band diagrams of unstrained and biaxially tensile strained Si.

5.2   1/*f* noise characteristics

The implementation of mechanical stress in the Si inversion layer in order to enhance the carrier mobilities makes the device fabrication more complex. The increased complexity may lead to unwanted effects such as

4. 1/f noise performance of advanced CMOS devices

increased LF noise. It has been reported that local stress produced by shallow-trench-isolation (STI)[96] as well as SiGe source/drain regions[97] can increase the LF noise, but there are also reports of unchanged behaviour (stress by Si_3N_4 cap).[97] Strained-Si MOSFETs fabricated on SiGe virtual substrates (global stress) could have severely increased oxide trap densities due to the defect-rich relaxed SiGe buffer. Presence of threading dislocations as well as outdiffusion of Ge from the SiGe substrate have been found to degrade the LF noise performance of such devices.[98-100] High temperature steps in the fabrication process should therefore be avoided. In contrast, Simoen and co-workers report reduced 1/f noise by a factor of two in strained Si devices fabricated on thin SiGe strained-relaxed buffer (SRB) layers.[33,101]

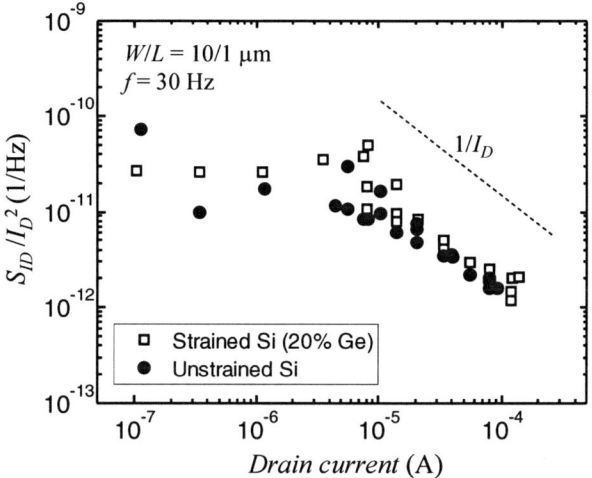

Figure 4-12. Normalized drain current noise plotted vs. drain current for tensile strained and unstrained Si nMOSFETs. $V_{DS} = -50$ mV.

The normalized drain current noise is plotted in Fig. 4-12 for strained-Si nMOSFET fabricated on 200 nm thin $Si_{0.8}Ge_{0.2}$ substrates. As seen, the 1/f noise performances of the strained and unstrained nMOSFETs are similar. Fig. 4-13(a) shows that the gate voltage noise also is unchanged in the strained-Si nMOSFETs in comparison with the unstrained references. The charge-pumping measurement in Fig. 4-13(b) indicates similar level of interface states implying that the gate oxide quality is conserved. Therefore, it is possible to optimize the mobility by using relaxed SiGe substrates to introduce strain in the Si channel without sacrificing the 1/f noise performance or gate oxide quality. The dominant 1/f noise mechanism was not easily determined in this particular case. The normalized drain current noise follows a $1/I_D$ behaviour above threshold according to Fig. 4-12, which could indicate that Hooge mobility noise is the origin of the $1/f$ noise. However, the frequency exponent strongly deviated from unity in weak inversion and at low V_{GT}. This observation indicates that trapping/release phenomena are present (as well), since $\gamma \sim 1$ is expected for Hooge mobility noise. Extraction of parameters for the number fluctuation noise model gives N_t around 4×10^{16} cm^{-3}eV^{-1} (both devices) and α equal to 2.3×10^4 Vs/C (unstrained) or 1.7×10^4 Vs/C (strained Si). The Hooge noise model can be employed successfully in strong inversion with $\alpha_H = 2.4 \times 10^{-5}$ (strained) or 2.8×10^{-5} (unstrained).

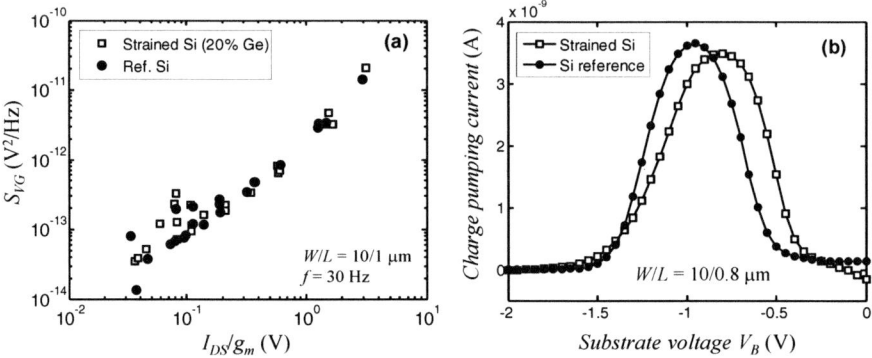

Figure 4-13. (a) Input gate voltage noise and (b) charge pumping current plotted for unstrained and tensile strained Si nMOSFETs.

However, if the fabrication process is not optimized or if too large amounts of stress are elaborated with, the outcome could be drastically elevated $1/f$ noise levels. Strained-Si n- and pMOSFETs were fabricated on a thick SiGe substrate containing 27% Ge resulting in an enhancement of the electron and hole mobilities by 120% and 55%, respectively (see Fig. 4-10). The gate voltage noise was increased by 1-2 orders of magnitude in the strained-Si nMOSFETs as shown in Fig. 4-14. The increased $1/f$ noise was attributed to partial strain relaxation and the presence of misfit dislocations. Interestingly, the strained-Si pMOSFETs exhibited lower $1/f$ noise than that in the reference unstrained Si pMOSFETs. The reason for the $1/f$ noise reduction is not clarified yet but the results again underscore the potential for these devices in high-speed analog applications thanks to their high mobility and low noise. In conjunction with these results, it should be mentioned that experiments where the channel orientation was shifted from the standard <110> to <100> were undertaken. The hole mobility was improved in the <100> direction, whereas the electron mobility was unchanged. The channel orientation was not found to have any significant influence on the 1/f noise performance neither in the strained-Si MOSFETs nor the unstrained ones.

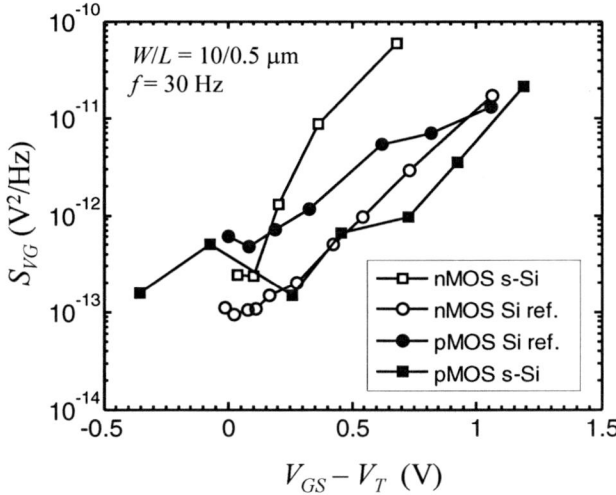

Figure 4-14. Input gate voltage noise studied vs. $V_{GS} - V_T$ for strained-Si (s-Si) and unstrained reference Si n- and pMOSFETs (note: axis scale $-V_{GS} + V_T$ for pMOS). The strained-Si devices were fabricated on a thick SiGe virtual substrate containing 27% Ge. $|V_{DS}| = 50$ mV.

6. SILICON-ON-INSULATOR DEVICES

6.1 Device structures and characteristics

The concept of fabricating a MOSFET in a thin silicon film situated on top of an insulating substrate, hence the name silicon-on-insulator (SOI), is nowadays an established technology which offers several advantages over the standard technology of fabricating MOSFET on bulk Si wafers. The SOI MOSFETs can be divided in three categories according to the Si body thickness or doping type; partially depleted (PD), fully depleted (FD) and accumulation mode devices (AM). Fully depleted SOI devices are fabricated on a thin Si body which is fully depleted. These devices are very attractive for future generations of ultra-scaled CMOS devices thanks to enhanced performance in terms of high speed and low power consumption as well as improved scalability.[102-107] The SOI MOSFET has a very small body effect coefficient m, around 1.05-1.1 in a FD device and equal to 1 in a PD device. It is readily observed from Eqs. (3-8) and (3-10) that a small m translates to a high drive current, high transconductance and small subthreshold slope (close to 60 mV/dec). In SOI devices with source/drain junctions reaching through the body, the parasitic source and drain capacitances are also decreased.

SOI devices were for a long time mainly used in harsh-environment electronics for military, space and high-energy physics applications that require high radiation hardness. The SOI substrates used to be expensive and of moderate quality, making it difficult for the SOI technology to compete with bulk Si. New fabrication technologies for SOI have markedly improved wafer quality. The revolutionary smart-cut technology[108] provides prime quality and relatively inexpensive UNIBOND substrates at the same time as the CMOS technology on bulk Si is facing a number of critical limitations. The performance benefits offered by the SOI technology have now made it very attractive for CMOS logic, memories and analog circuits.[102,103,109]

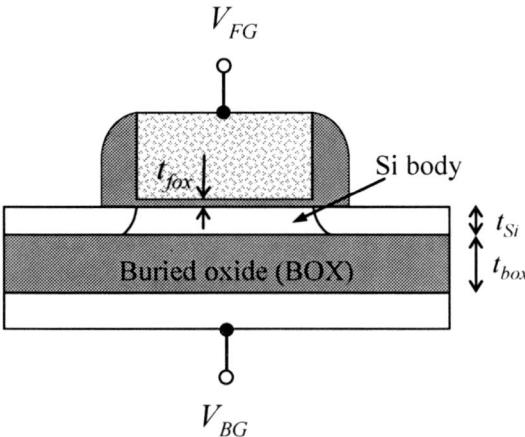

Figure 4-15. Schematic cross section of a SOI MOSFET.

The different types of SOI devices will be discussed in more detail in the next subsections. Fig. 4-15 shows a schematic cross section of a SOI MOSFET where some notations and dimensions that we will use later are defined. Band diagrams for each of the three SOI devices are presented in Fig. 4-16.

4. 1/f noise performance of advanced CMOS devices

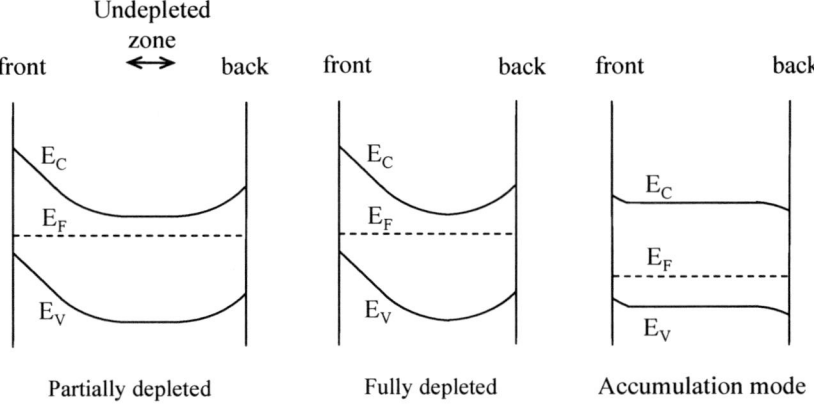

Figure 4-16. Schematic band diagrams for PD, FD and AM SOI pMOSFETs.

6.1.1 Partially depleted (PD) SOI

Partially depleted devices are fabricated on SOI substrates with a thick body where

$$t_{Si} > 2\sqrt{4\varepsilon_{Si}\psi_B / qN_{body}} \; . \tag{4-13}$$

Only a part of the body is depleted and a neutral piece of Si exists. This type of device behaves exactly as a bulk MOS device with the exception of parasitic effects related to the electrically floating body. If the body is not connected to ground, charging of the body can occur due to impact ionization effects in the high-field region near the drain or tunneling of carriers through the gate oxide. The charging of the body will increase the body potential, which leads to a reduction of the threshold voltage and a an increase of the drain current (a "kink" in the I_D-V_{DS} or I_D-V_{GS} curves).

6.1.2 Fully depleted (FD) SOI

In fully depleted devices, the thin Si-body is fully depleted,

$$t_{Si} < \sqrt{4\varepsilon_{Si}\psi_B / qN_{body}} \; . \tag{4-14}$$

The depletion region covers the whole body and does not extend with gate bias. Fully depleted SOI devices are normally free (or almost free) of floating body effects, but can appear if the back interface is in accumulation.[110] The short-channel effect can effectively be controlled by making the Si body ultra-thin, approximately less than 1/3 of the gate length.[111] The body can actually be left undoped if it is sufficiently thin, which eliminates the Coulomb scattering from ionized impurities and lowers the effective electric field, hence resulting in improved mobility.

Fully depleted MOSFETs are in theory dual-gate devices, the inversion channel can be controlled by both the front and the back gate voltage. A multiple-gate architecture provides an even better electro-static control of the channel by the gate, and consequently even better scalability of the technology. However, the buried oxide in a standard SOI wafer is usually much thicker than the front gate oxide. To obtain a symmetric behaviour, the back gate voltage must be much larger than the voltage on the front ($V_{BG} \sim t_{box}/t_{fox} \cdot V_{FG}$). SOI substrates are used as the starting material to fabricate multiple-gate devices such as double-gate SOI MOSFETs, FinFETs, Omega FETs, and Gate-all-around MOSFETs. Devices with three (e.g. Omega FET) or four gates (e.g. gate-all-around FET) are attractive solutions in order to be able to scale the devices with maintained control of the channel to the extremely short gate lengths (~ 10 nm) at the end of the ITRS roadmap. We will discuss devices with multiple gates further in section 9.

6.1.3 Accumulation mode (AM) SOI

A special type of FD SOI MOSFET is realized if the Si body is doped p-type for a pMOSFET and n-type for an nMOSFET, respectively. By using a thin Si-body, the body is fully depleted in the off-state. In such case, no current flows between source and drain as well as the short-channel effects are well controlled. In the on-state, the depletion region gradually disappears and an accumulation channel is formed. One important difference compared to standard FD devices is that the inversion charge is spread out in the body instead of piling up the front and back oxide interfaces,[112] see the schematic band diagram in Fig. 4-16. The conduction path is therefore partly buried in the Si body, which results in enhanced mobility as well as lower $1/f$ noise in the AM SOI devices. SOI technology is therefore useful in order to realize buried channel devices.

6.2 $1/f$ noise characteristics

In the past, MOS transistors fabricated on SOI substrates were notorious for their poor LF noise performance compared to bulk CMOS. There are several reasons why SOI MOSFETs were noisier; in addition to the noise generated at the front oxide interface, appreciable LF noise can also be generated at the back interface, from defects in the Si-body, and due to

4. 1/f noise performance of advanced CMOS devices

floating body effects.[113] Naturally, the type of device (FD, PD or AM SOI) and choice of SOI substrate (UNIBOND, SIMOX etc) can have a large impact on the LF noise performance. Extensive advances have recently been made in improving the quality of the SOI substrates providing prime quality and relatively inexpensive UNIBOND substrates for example, which potentially can result in improved noise performance in SOI transistors. However, by comparing the recently reported trap densities for SOI MOSFETs, most of them fabricated on UNIBOND substrates, with those for standard bulk CMOS in Fig. 4-3, the values for SOI are still noticeable higher.

The other side of the coin is that SOI MOSFETs can be designed to produce very low $1/f$ noise under certain conditions.[42,114,115] The position of the conduction channel can be varied from surface to bulk mode by the front and back gate voltages, which affects the noise properties. Ultimately, for sufficiently thin Si-body thickness (below ~10 nm) the interior of the body is inverted which separates the carriers from the noisy oxide interfaces. The concept of *volume inversion*, first invented by Balestra et al.,[116] is very attractive to achieve high mobility, transconductance and low noise. In the following subsections, we will discuss the characteristic $1/f$ noise properties of the PD, FD and AM SOI devices in more detail.

6.2.1 Partially depleted SOI

As mentioned earlier, PD MOSFETs are prone to floating body effects. Closely linked to the static behaviour with a shift of the threshold voltage and a kink in the drain current under certain conditions, a noise overshoot is found to occur at the same bias conditions. The kink-related excess low-frequency noise appears as Lorentzian-like components in the spectra and arises from the shot noise in current that discharge the body through the S/B junction.[117] It is typical that the noise plateau ($\propto 1/I_{SB}$) and the corner frequency ($\propto I_{SB}$) of the Lorentzian-like excess noise shift with bias. The LF noise observed at a certain frequency displays a sharp maximum for a particular V_{DS} (or V_{GS}) that can be one or two orders of magnitude above the normal noise level. Note that the impact ionization rate for holes is smaller than that for electrons, resulting in a higher kink onset drain-source voltage in pMOS SOI devices. The kink-related excess noise can be almost eliminated by connecting the body to the ground.[27,117]

6.2.2 Fully depleted SOI

A FD MOSFETs can be controlled by both the voltage at the front gate (V_{FG}) and the voltage at the back gate (V_{BG}). This coupling between the front and back gates leads to increased 1/f noise as the noise generated from traps at the back interface then couples to the output and adds to the drain current noise. The following equation describes the coupling effect on the drain current noise[28]

$$S_{I_D} \approx g_{m,f}^2 \left(S_{Vfb,f} + \frac{C_{Si}^2 \cdot C_{box}^2}{C_{fox}^2 (C_{box} + C_{Si})^2} S_{Vfb,b} \right) \quad (4\text{-}15)$$

where $g_{m,f}$ is the front-channel transconductance. $S_{Vfb,f}$ and $S_{Vfb,b}$ are the front and back gate flat-band voltage PSDs, respectively. The Si body capacitance and the buried oxide capacitance equal $C_{Si} = \varepsilon_{Si}/t_{Si}$ and $C_{box} = \varepsilon_{ox}/t_{box}$, respectively. At the limit $C_{Si} \gg C_{box}$

$$S_{I_D} \approx g_{m,f}^2 S_{Vfb,f} (1 + N_{t,b}/N_{t,f}). \quad (4\text{-}16)$$

Here, $N_{t,b}$ and $N_{t,f}$ are the oxide trap densities for the back and front oxide, respectively. Obviously, if the buried oxide is of poor quality ($N_{t,b} \gg N_{t,f}$), the 1/f noise performance is severely degraded. For the Hooge mobility fluctuations, the following relationship can be predicted (uncorrelated noise sources)

$$\frac{S_{I_D}}{I_D^2} = \frac{q\alpha_{H,f}}{fWLQ_{i,f}} \frac{I_{D,f}^2}{I_D^2} + \frac{q\alpha_{H,b}}{fWLQ_{i,b}} \frac{I_{D,b}^2}{I_D^2}. \quad (4\text{-}17)$$

The subscripts refer to front (f) and back channel (b) respectively. When only a small current is carried at the back interface, mobility fluctuation noise will mainly be generated in the front channel. The number fluctuation noise is more sensitive to coupling effects due to the $1/C^2$ dependence compared to the $1/C$ scaling for mobility fluctuations.

The LF noise overshoot due to floating body effects is greatly suppressed in FD devices, but is not always completely eliminated.[25,117] Especially if the back gate is biased in accumulation, kink-related excess noise can cause problems in FD devices as well.[28]

6.2.3 Accumulation mode SOI

In this subsection, we will examine the 1/f noise properties of AM SOI pMOSFETs and demonstrate that improved low-frequency noise performance can be achieved by exploiting the buried channel concept.

4. 1/f noise performance of advanced CMOS devices

The investigated SOI pMOSFETs were fabricated on a 20 nm thin Si-body with a light p-type doping ($0.6\text{-}1 \times 10^{15}$ cm^{-3}). The devices exhibited a nearly ideal subthreshold slope of 62 mV/dec and well-controlled short-channel effects down to 0.1 μm gate length. The hole mobility showed a peak value around 130 cm^2/Vs which is significantly higher than in comparable bulk Si pMOSFETs. The average distance of the inversion charge from the front oxide interface was simulated for the AM SOI pMOSFETs using the following parameters: $t_{ox} = 3$ nm, $t_{box} = 400$ nm, $N_A = 1 \times 10^{15}$ cm^{-3}. The results are shown in Fig. 4-17 and compared with corresponding simulation results for a bulk Si pMOSFET designed for ~0.1 μm gate length ($t_{ox} = 3$ nm, $N_D = 7.7 \times 10^{17}$ cm^{-3}). As seen in Fig. 4-17(a), the carriers are further away from the front oxide interface in the AM SOI devices. Fig. 4-17(b) displays the hole distribution in the Si body, the holes are spread out over the entire body when $V_{FG} = V_T$ but piles up at the front oxide interface at higher (front) gate voltage overdrives.

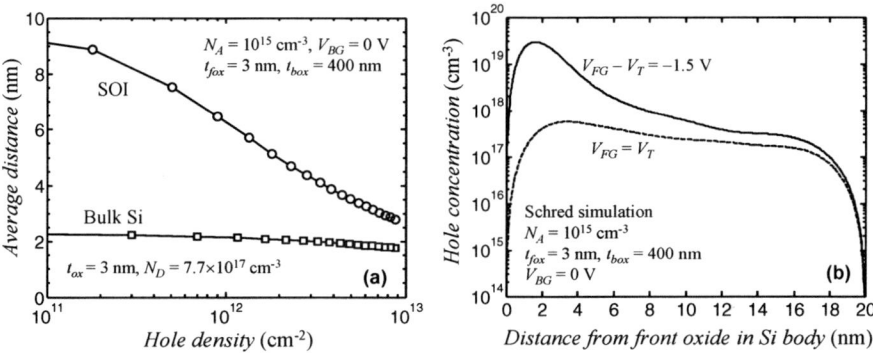

Figure 4-17. (a) TCAD simulation of the average distance of the inversion carriers from the front gate oxide interface in a SOI and a bulk Si pMOSFET. (b) Simulated distribution of holes in the 20-nm thin body of a SOI pMOSFET. Schred[118] was used for both TCAD simulations.

The normalized drain current noise for these two devices, both fabricated in a similar process with 3-nm pure thermally grown SiO$_2$ as gate dielectrics, is plotted in Fig. 4-18. A clear noise reduction is found for the SOI pMOSFETs and a low $\alpha_H \sim 9 \times 10^{-6}$ was extracted. This is explained by the buried conduction path in the AM devices, which separates the carriers from the gate oxide interface. Note that higher 1/f noise often is found in FD SOI devices in comparison with Si devices due to the fact that the number fluctuation noise increases with the factor $(1+N_{t,b}/N_{t,f})$. Here, it was found that number fluctuation noise may contribute appreciably in the SOI devices when they were operated close to the threshold. In strong inversion, on the other hand, mobility fluctuation noise took over as the dominant $1/f$ noise mechanism. According to our analysis in chapter 3.4.3.3, a lower value of α_H can be expected for the mobility fluctuation noise in a buried channel device.

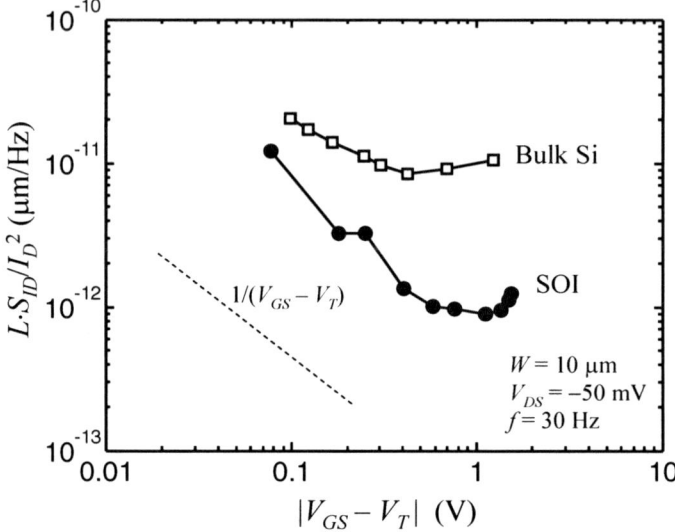

Figure 4-18. Drain current noise measured in a SOI and a bulk Si pMOSFET, both with 3-nm SiO_2 as (front) gate oxide.

The utilization of a compressively strained SiGe channel on SOI has been found to enhance the hole mobility by more than 60%.[119] The $1/f$ noise performance of these devices was, however, similar to the Si channel ones since the Si-cap was too thin (< 1 nm) to effectively separate the inversion carriers from the oxide interface.[42] An optimization of the Si-cap thickness is necessary for improved low-frequency noise performance in buried SiGe channel pMOSFETs, in line with our conclusions in section 4. No evidence of kink-related excess noise was found either for the Si or the SiGe device in the studied bias range as seen in Fig. 4-19, indicating that the problem with floating body is possible to eliminate (or pushed to higher drain bias as suggested by Tseng *et al.*[117]) in the FD device architecture. In Fig. 4-19, the dotted curve shows a typical appearance of kink-related excess noise. The exact drain bias where the overshoot occurs is related to gate bias, frequency as well as technology issues such as abruptness of the doping profile.

4. 1/f noise performance of advanced CMOS devices

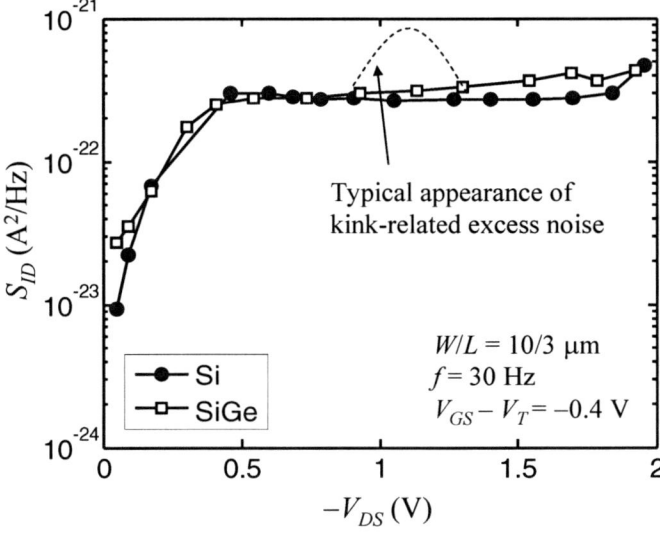

Figure 4-19. Drain current noise vs. drain-source voltage for two AM SOI pMOSFETs.

6.2.4 Schottky Barrier (SB) MOSFETs

The S/D resistance is becoming a difficult problem in ultra scaled CMOS devices, especially if they are fabricated on ultra-thin body SOI substrates. An attractive solution to lower the S/D resistance is by forming Schottky barrier contacts in the S/D regions, usually implemented with metal silicides. High performance Schottky Barrier (SB) pMOSFETs with sub-30-nm gate lengths showing a f_T = 280 GHz have recently been reported.[120] These devices were fabricated with platinum silicide contacts in the S/D regions.

We have studied SOI pMOSFETs where nickel silicide (NiSi) was formed in the S/D. SB MOSFET behaviour was achieved when the NiSi from the S/D regions penetrated into the channel and formed NiSi-Si Schottky junctions as shown in Fig. 4-20(a).[121] The I_D-V_{GS} characteristics of the SB pMOSFET and a reference device where the Schottky barrier is formed in the extension region are compared in Fig. 4-20(b). The drain current in the SB device is limited by the reverse biased Schottky barrier at the source side at lower bias. The width of the barrier is decreased at higher gate bias, which enhances the tunneling current across the barrier and the S/D resistance reduces. Fig. 4-21 emphasizes the importance of a low S/D resistance also for the LF noise performance.

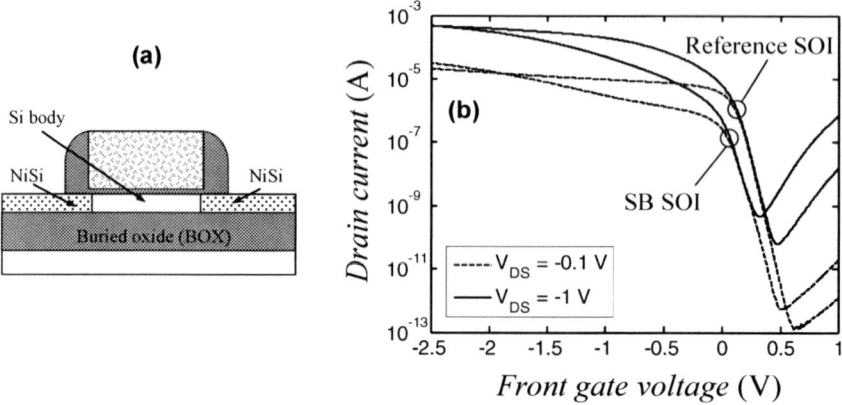

Figure 4-20. (a) Schematic structure and (b) I_D-V_{GS} characteristics of the Schottky Barrier (SB) pMOSFETs. $W/L = 10/1$ μm.

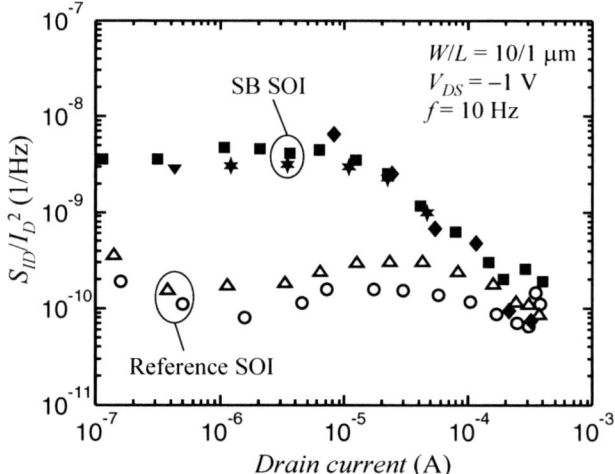

Figure 4-21. Normalized drain current noise plotted for several NiSi Schottky Barrier and reference SOI pMOSFETs. Note that the reference devices in this experiment showed high S/D resistances, explaining the unusual bias dependence of the normalized drain current noise (from von Haartman et al.[42]).

As seen in Fig. 4-21, the normalized drain current noise is independent of I_D, which indicates that the noise originates from the source side in both cases. This follows from Eq. (3-12) by setting $g_{ch} = 0$ (saturation). The fact that the contacts are poor and their area is small ($t_{si} \times W = 0.02 \times 10$ μm^2) make them very noisy. The normalized drain current noise starts to decrease and approach the reference device at $I_D \sim 10^{-5}$ A, which indicates that the noise properties of the contact improve as the contact resistance of the NiSi-Si junction starts to decrease. Note that the reference SOI device in Fig. 4-21 shows an unusual $1/f$ noise behaviour compared to standard MOSFETs due to the large S/D resistance. An improved design of the SB MOSFET together with a transition to a material with low barrier (such as PtSi) is expected to yield reduced S/D resistances as well as decreased $1/f$ noise. This particular example illustrates how sensitive the $1/f$ noise is to the presence of bottlenecks for the current transport.

7. MOSFETS WITH HIGH-K GATE DIELECTRICS

7.1 Replacement of SiO$_2$ with high-k materials

As the gate length is scaled down, the gate oxide capacitance needs to increase in order to control the short channel effects and fulfil the scaling requirements. This evolution has pushed the thickness of the SiO$_2$ gate oxide to its physical limits. As of today (2006), the high performance logic devices have a SiO$_2$ thickness of around 1.1 nm. The gate tunneling current increases exponentially with decreasing thickness.[122] For oxide thicknesses around 1-1.5 nm, depending on application, the gate leakage current becomes intolerably high.[5] A high gate leakage current causes problems such as increased standby power consumption, deteriorated reliability and device lifetime, and can ruin the whole device operation. By replacing the SiO$_2$, which has a dielectric constant k of 3.9, with a material with higher dielectric constant, the gate leakage current can be maintained at a low level. This is because a physically thicker gate dielectric can be used by replacing the SiO$_2$ with a so-called high-k material.[123,124] The equivalent oxide thickness (EOT) is defined as

$$t_{EOT} = \frac{k_{SiO_2}}{k_{high-k}} t_{high-k} \qquad (4\text{-}18)$$

which corresponds to the equivalent thickness of SiO$_2$ giving the same capacitance as of the high-k gate dielectric with thickness t_{high-k} and dielectric constant k_{high-k}.

One of the reasons for the outstanding achievements with CMOS technology is that an excellent insulator, SiO$_2$, has been available. To replace the silicon dioxide is therefore an enormous challenge. The high-k materials that are to replace the SiO_2 must satisfy various requirements namely: (i) thermodynamically stable together with Si, (ii) process compatible with CMOS, (iii) negligible interface layer formation, (iv) sufficient band offsets to act as tunneling barriers for electrons and holes, (v) high quality interface with Si and (vi) low defect densities.[125] A large number of high-k materials, listed in Table 4-2, have been researched in combination with CMOS and several difficulties have been encountered.

Table 4-2. Properties of the most commonly researched high-k dielectric materials, from Wilk, Wallace and Anthony.[123] Note that the ΔE_V values were calculated from the Band gap and ΔE_C values in the table according to $\Delta E_V = E_{g,high-k} - \Delta E_C - E_{g,Si}$. The properties of mixed materials such as $HfSiO_xN_y$ and $HfAlO_x$ depend on the content of the different atoms in the compound. The dielectric constants of $HfSiO_xN_y$ and $HfAlO_x$ are reported to range between 9-14 and 14-20, respectively.

Material	Dielectric constant k	Band gap E_g (eV)	ΔE_C to Si (eV)	ΔE_V to Si (eV)
SiO_2	3.9	8.9	3.2	4.6
Si_3N_4	7	5.1	2	2
Al_2O_3	9	8.7	2.8	4.8
Y_2O_3	15	5.6	2.3	2.2
La_2O_3	30	4.3	2.3	0.9
Ta_2O_5	26	4.5	1-1.5	1.9-2.4
TiO_2	80	3.5	1.2	1.2
HfO_2	25	5.7	1.5	3.1
ZrO_2	25	7.8	1.4	5.3

From noise performance point of view it is worrying with the reports about degraded mobility,[124,126-129] and high density of traps and fixed charges.[123,124,130,131] Not surprising, most reports so far indicate 1-3 orders of magnitude higher $1/f$ noise compared to CMOS devices with thermal SiO_2.[12,13,41,43,44,132-137] Other problems that have been frequently observed include threshold voltage instabilities,[138,139] dopant penetration, crystallization upon heating, as well as points (i) to (vi) above. Due to these problems, the semiconductor industry has postponed the introduction of high-k materials and instead used existing technology with some modifications. By adding nitrogen to silicon dioxide, forming so called oxynitrides (SiO_xN_y), the dielectric constant is increased in proportion to the nitrogen content up to a value of 7. Oxynitrides also have the important advantage of suppressing boron penetration from a p^+ doped poly-Si gate and improving hot-carrier reliability.[54,140] The use of oxynitride is a short term solution until some high-k gate dielectric integrated in CMOS technology is ready for mass production. Hafniumoxide (HfO_2) with a dielectric constant of 20-25 is the most studied high-k material and the

4. 1/f noise performance of advanced CMOS devices

leading contender to replace oxynitrides. Hafniumsilicates (HfSiON) might be an intermediate solution since they are more resistant to crystallization and presently have lower defect densities than HfO$_2$.[131]

Extensive experimental studies of pMOSFETs with high-k gate dielectrics have been performed at the device technology laboratory, KTH, Kista. The most important results and findings from this work will be presented and analyzed here. The high-k gate dielectric stacks researched in our work were, with one exception, deposited by means of Atomic Layer Deposition (ALD), performed at ASM Microchemistry Oy, Finland. Other techniques, such as Physical Vapour Deposition (PVD, for example sputtering and evaporation), molecular beam epitaxy (MBE), and Metal Organic Chemical Vapour Deposition (MOCVD) are available as well. There exists no systematic study of the impact of the deposition methods in relation to noise, as far as we know. However, ALD is recognized for providing uniform layers with low defect densities, and is together with MOCVD the most frequently used deposition technique for high-performance transistors. Claeys *et al.* reported that N_t was found to be lower in ALD high-k layers than that for MOCVD.[141]

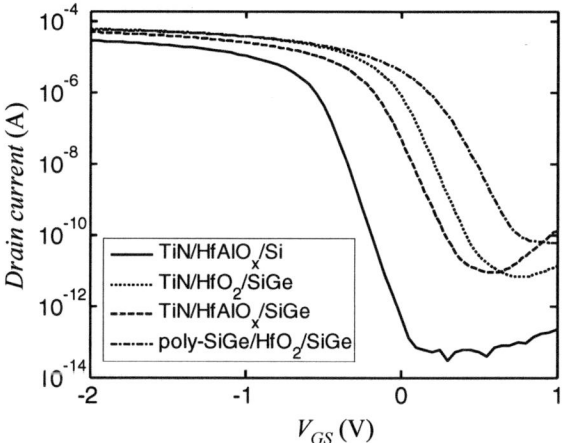

Figure 4-22. I_D-V_{GS} characteristics for Si and Si$_{0.7}$Ge$_{0.3}$ pMOSFETs with HfO$_2$ based high-k gate dielectrics and a TiN or poly-SiGe gate. W/L = 10/1 µm, V_{DS} = −50 mV. The 5-nm thick gate dielectric stack was grown by ALD with 0.5 nm Al$_2$O$_3$ at top and bottom.

P-channel transistors were fabricated with high-k stacks consisting of Al$_2$O$_3$ layers at top and bottom, sandwiching an HfAlO$_x$, HfO$_2$ or Al$_2$O$_3$ layer in the middle. ALD TiN or *in-situ* p$^+$ doped poly-SiGe was used as gate electrode material.[43,142] Fig. 4-22 shows I_D-V_{GS} curves for surface Si and SiGe channel pMOSFETs with HfO$_2$ based high-k gate dielectrics. The Si device shows a low subthreshold slope of ~75 mV/dec and an interface state density of around 5×10^{11} cm^{-2}eV^{-1}. The hole mobility is slightly reduced compared to the universal mobility curve, as seen in Fig. 4-23(a). It is commonly observed that MOSFETs with high-k gate dielectrics show a, to various extent, degraded mobility compared to the universal mobility curves that are derived for Si MOSFETs with SiO$_2$ gate oxide. Scattering from charges in the high-k layer or at the interfaces, remote phonon scattering, remote surface roughness scattering, and crystallization are some of the possible sources behind the mobility degradation.[127]

For the Si transistor, the origin of the lower mobility is ascribed to remote phonon scattering. The "soft" bonds in a highly polarizable material are associated with low-energy ("soft") optical phonons giving rise to additional scattering of the carriers in the remote inversion layer.[143] This scattering source does not play a major role in SiO$_2$ due to the stiff bond and low dielectric constant, but reduces the mobility in devices with high-k gate dielectrics roughly in proportion the value of the dielectric constant. The hole mobility in the surface SiGe channel devices is enhanced compared to Si, but suffers from Coulomb scattering from fixed charges at low electric fields. An analysis of the temperature sensitivity factor of the mobility, $d(1/\mu_{eff})/dT$, can be used to determine the dominant scattering mechanisms. Fig. 4-23(b) shows the temperature sensitivity factor of the mobility for the high-k MOSFETs. Phonon scattering gives a positive temperature sensitivity factor (phonon limited mobility decreases with temperature), whereas Coulomb scattering gives a negative temperature sensitivity factor. Surface roughness scattering increases weakly with temperature, the temperature sensitivity factor is therefore expected to be small but positive when this mechanism is dominant. From Fig. 4-23(b) it can be observed that the high-k/Si device shows a higher temperature sensitivity factor than that for the SiO$_2$/Si device, indicating influence of remote phonon scattering in the former device. The temperature sensitivity factor is reduced for the SiGe devices due to Coulomb scattering from trapped and fixed charges as indicated earlier.

4. 1/f noise performance of advanced CMOS devices

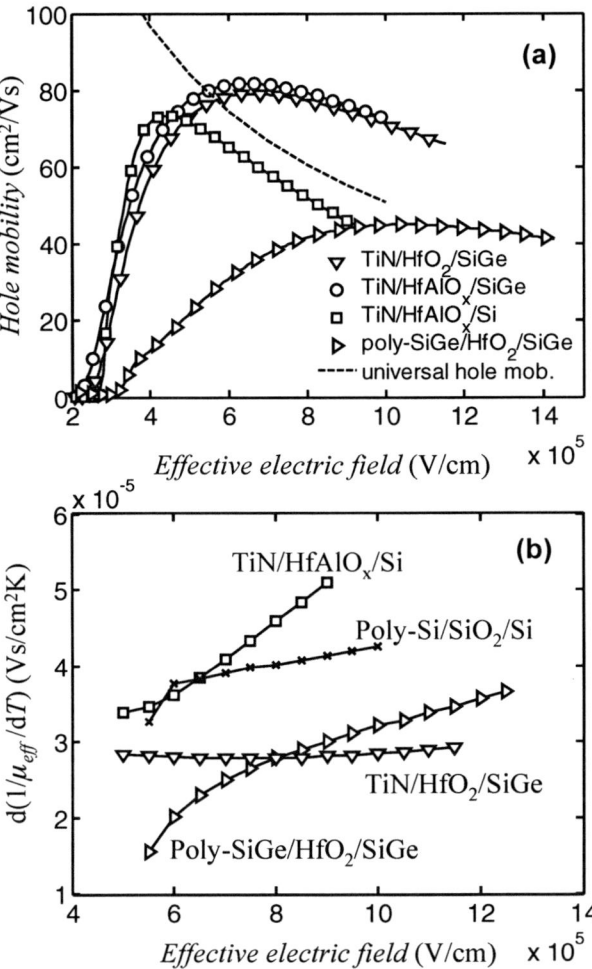

Figure 4-23. (a) Effective hole mobility and (b) temperature sensitivity factor of the mobility plotted versus effective electric field for Si and $Si_{0.7}Ge_{0.3}$ pMOSFETs with HfO_2 based high-k gate dielectrics and a TiN or poly-SiGe gate. The temperature sensitivity factor for the $TiN/HfAlO_x/SiGe$ device is almost on top of the curve for the $TiN/HfO_2/SiGe$ device (not shown). Reprinted from von Haartman et al.[41] with permission from IEEE.

7.2 1/f noise characteristics

The replacement of the SiO_xN_y gate dielectrics with materials having a higher dielectric constant k is required for future CMOS technologies beyond the 65 nm node, in order to maintain a low gate leakage current at the same time as the gate oxide capacitance is scaled up. From a noise perspective, this technology shift leads to orders of magnitude (1-3) higher

$1/f$ noise compared to CMOS devices with thermal SiO_2. The higher $1/f$ noise is in most cases ascribed to a high density of traps in the high-k gate dielectrics. However, as will be shown here, Hooge mobility fluctuation noise is also important, especially in p-channel MOSFETs. Traps in the high-k material, located from near the channel interface to several nm inside the bulk of the material, can contribute to the $1/f$ noise. Interfaces between different materials are notorious for high trap densities and can cause g-r noise bumps. Simoen and co-workers have demonstrated that electrons tunneling to and from traps in an HfO_2 layer deposited on 2.1-nm SiO_2 is the origin of the $1/f^\gamma$ noise in their devices, which illustrates the McWorther type noise mechanism.[144] This agrees with the observation of instabilities in the threshold voltage, which has been explained by charging and discharging of traps in the high-k material by tunneling.[138]

The oxide trap density N_t in the high-k materials ranges between 4×10^{17} and 1×10^{20} $cm^{-3}eV^{-1}$ as reported from LF noise characterizations. In Fig. 4-24, the different high-k materials are compared, the oxide trap density is shown in (a) and the Hooge parameter in (b). Extracted trap densities for nitrided SiO_2 range between 1×10^{16} and 1×10^{18} $cm^{-3}eV^{-1}$, as shown in Fig. 4-3. Trap-density profiles in HfO_2 and Al_2O_3 gate dielectrics derived from various charge-pumping schemes are consistent with the results in Fig. 4-24(a).[130,145] It has also been reported that the trap densities in SiO_2 increase when high-k materials are deposited on top.[12,13] The values at large t_{EOT} in Fig. 4-24 are in most cases for devices with a thick layer of SiO_2 between the high-k layer and the substrate, which explain why these devices perform better. The Hooge parameter is found to be in the range 10^{-4} and 10^{-2} for the transistors with high-k gate dielectrics, which is higher than in conventional MOSFETs ($\alpha_H \sim 10^{-6} - 10^{-3}$). The extracted N_t and α_H values from the work performed by KTH[41-46] are included in Fig. 4-24 along with results published in the literature from other groups.[12-14,24,26,31,36-39,132,134-137] Table 4-3 presents a more detailed overview of extracted device and noise parameters in this work. In the following subsections, the LF noise properties of MOSFETs with high-k gate dielectrics are discussed in more detail.

4. 1/f noise performance of advanced CMOS devices

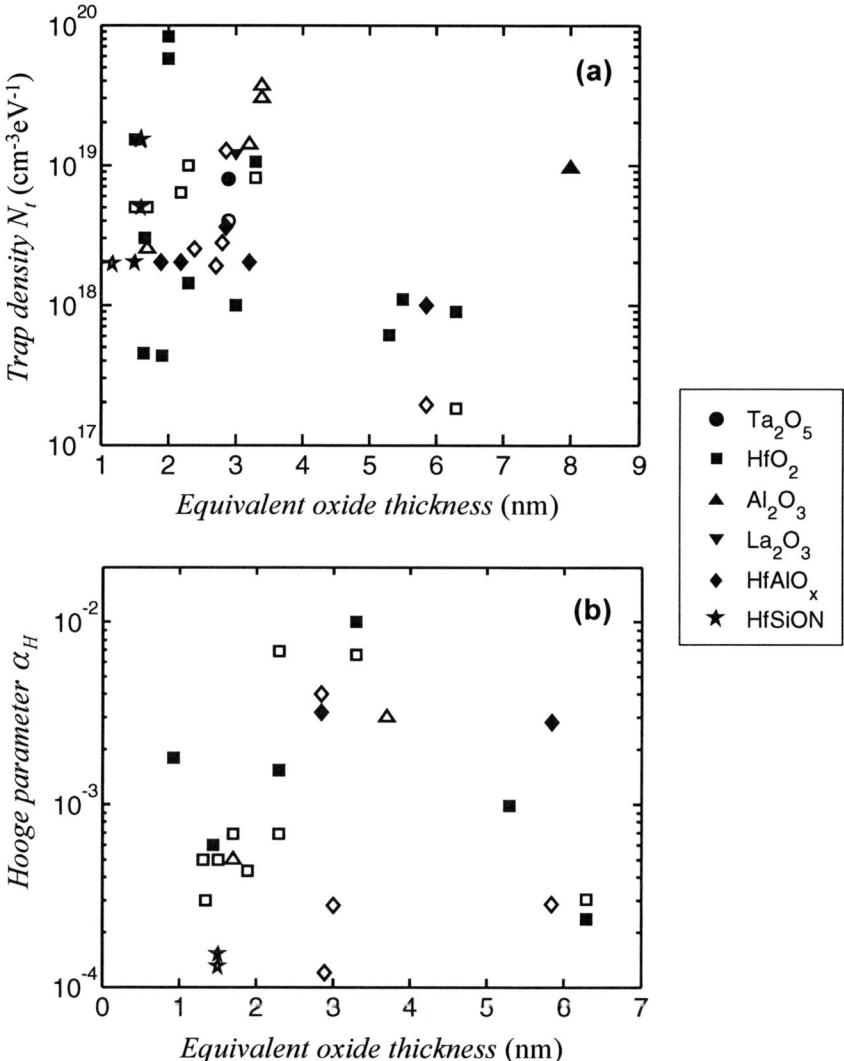

Figure 4-24. Summary of reported values (including this work) on (a) the oxide trap density N_t and (b) the Hooge parameter α_H for different high-k materials. Filled symbols denote nMOS and open symbols pMOS devices.

Table 4-3. Summary of extracted device and noise parameters for the pMOSFETs with high-k gate dielectrics studied in this work: threshold voltage (V_T), subthreshold slope (SS), equivalent oxide thickness (t_{EOT}), density of interface states (D_{it}), gate-leakage current density (J_G), oxide charge density (N_{ox}) at flat band, oxide trap density (N_t), and Hooge parameter (α_H). From von Haartman, Malm, and Östling.[41]

V_T (V)	SS (mV/dec)	t_{EOT} (nm)	D_{it} (cm^{-2}eV^{-1})	J_G @ V_G = −1.5 V (A/cm^2)	N_{ox} (cm^{-2})	N_t (cm^{-3}eV^{-1})	α_H @ V_{GT} = 0.1V
Gate: TiN, High-k: Al$_2$O$_3$/HfAlO$_x$/Al$_2$O$_3$ (0.5/4/0.5 nm), Channel: Si							
−0.54	75	3.0	4.8×10^{11}	2.2×10^{-7}	−5×10^{11}	1.9×10^{18}	4.1×10^{-4}
Gate: TiN, High-k: Al$_2$O$_3$/HfAlO$_x$/Al$_2$O$_3$ (0.5/4/0.5 nm), Channel: Si$_{0.7}$Ge$_{0.3}$							
−0.21	110	2.9	3.5×10^{12}	1.3×10^{-7}	−1×10^{12}	1.7×10^{18}	2.2×10^{-4}
Gate: TiN, High-k: Al$_2$O$_3$/HfO$_2$/Al$_2$O$_3$ (0.5/4/0.5 nm), Channel: Si$_{0.7}$Ge$_{0.3}$							
−0.07	100	2.3	3.2×10^{12}	0.9×10^{-7}	−2×10^{12}	6.5×10^{18}	6.9×10^{-4}
Gate: TiN, High-k: Al$_2$O$_3$ (5 nm), Channel: Si$_{0.8}$Ge$_{0.2}$							
0.19	140	3.7	3.0×10^{12}	3.5×10^{-6}	−5×10^{12}	2.4×10^{19}	3.0×10^{-3}
Gate: poly-SiGe, High-k: Al$_2$O$_3$/HfO$_2$/Al$_2$O$_3$ (0.5/3/0.5 nm), Channel: Si$_{0.7}$Ge$_{0.3}$							
0.05	100	1.6	3.7×10^{12}	5.1×10^{-6}	−5×10^{12}	9.0×10^{18}	6.8×10^{-4}

7.2.1 Experiment and theory

One particular problem with performing LF noise measurements on transistors with high-k gate dielectrics is the threshold voltage instability. A LF noise measurement from 1 Hz to 100 Hz typically takes several minutes. During this time period, the threshold voltage can shift a few tenths of volts, in the worst case. As the threshold voltage is not fixed, care must be taken when studying the noise variation with the gate voltage overdrive for example. In our measurements, the devices were given some time to settle after each bias point adjustment. The drain current was measured before and after the noise measurements at each bias point and the average current was used in the calculations. The drift and variations in the average drain current and transconductance were acceptably low (< 1%) in most cases, except at very low currents in the subthreshold region. For that reason, noise measurements below I_D =100 nA in an L = 1 µm MOSFET were not found to be useful, since I_D could vary with more than 10%.

From theoretical viewpoint, the difference between MOSFETs with high-k or SiO$_2$ gate dielectrics concerns the tunneling attenuation length λ. The barrier height and the effective mass of the carriers differ for the high-k materials and SiO$_2$, see Table 4-2 for barrier heights of the most common high-k materials on Si. Min et al. calculated λ values for HfO$_2$ and Al$_2$O$_3$ in case of electrons tunneling from the conduction band in the Si to the high-k gate dielectrics.[13] Table 4-4 summarizes the λ values for the HfO$_2$, Al$_2$O$_3$ and SiO$_2$ gate dielectrics. As noise magnitude differences of a factor of two

4. 1/f noise performance of advanced CMOS devices 145

are considered as small, the differences in λ from Table 4-4 can also be considered as small.

Table 4-4. Tunneling attenuation length λ calculated for SiO_2, HfO_2 and Al_2O_3 on Si.

Material	λ (cm) electron tunneling from Si CB	λ (cm) hole tunneling from Si VB
SiO_2	1.0×10^{-8}	0.81×10^{-8}
HfO_2	2.1×10^{-8}	-
Al_2O_3	1.1×10^{-8}	-

A complicating circumstance is if the high-k stack is composed of several layers of different materials. Usually, a thin SiO_2 interfacial layer is present between the substrate and the high-k stack that may not be intentionally grown. In such cases, the calculation of λ is more complicated.[146] Our devices have gate stacks composed by two 5-Å Al_2O_3 layers sandwiching an HfO_2, $HfAlO_x$ or Al_2O_3 layer in the middle. Moreover, a 0-10 Å thick interfacial layer was found to be present between the bottom Al_2O_3 layer and the substrate. For the sake of simplicity and the insignificant differences in the λ parameter values, we used the values calculated for the SiO_2/Si(Ge) system in our work. Thus, for a $Si_{0.7}Ge_{0.3}$ channel λ decreases slightly to 0.79×10^{-8} cm for hole tunneling.

7.2.2 Dependence on gate dielectric material

Here, we will discuss the LF noise characteristics of different high-k materials in more detail. Published results are today available primarily for HfO_2, Al_2O_3, $HfAlO_x$ and HfSiON gate dielectrics. As mentioned earlier, HfO_2 is presently the main contender to replace oxynitrides in future CMOS devices and is by far the most studied high-k material. Additionally, a few LF noise studies of La_2O_3 and Ta_2O_5 are also available. A noise performance overview of the aforementioned materials was presented in Fig. 4-24.

Fig. 4-25 displays the normalized drain current noise extracted at 10 Hz versus drain current for pMOSFETs with Al_2O_3, $HfAlO_x$ and HfO_2 gate dielectrics. Details of the dielectric structures as well as extracted parameters of these devices were summarized in Table 4-3. The 1/f noise for the high-k pMOSFETs is 1-3 orders of magnitude higher than that for the SiO_2/Si reference device. On the positive side, the difference decreases with increasing bias down to around a factor of two or three at $I_D > 30$ μA. At lower bias, $HfAlO_x$ exhibits the lowest 1/f noise among the high-k materials, whereas the difference between them is small at high bias. The device with 5-nm Al_2O_3 is noisiest, but in another experiment (see next subsection) a device with 2-nm thick Al_2O_3 at the bottom interface was somewhat less

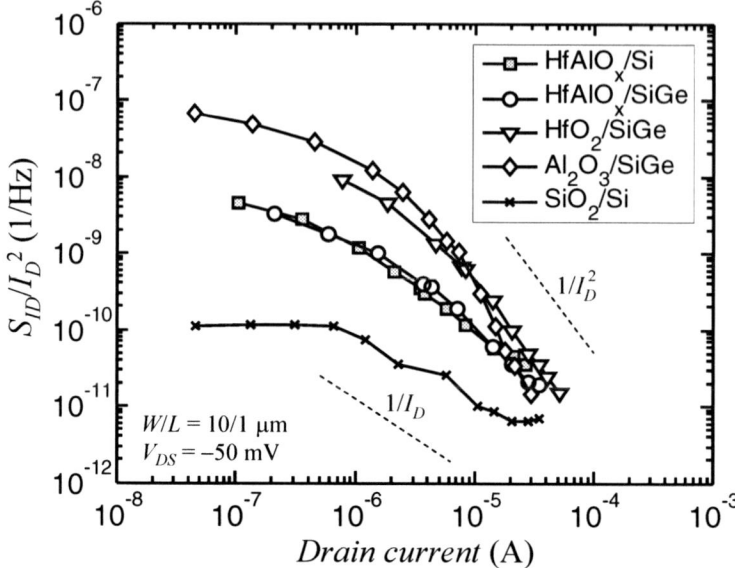

Figure 4-25. Normalized drain current noise at 10 Hz plotted vs. drain current for Si and SiGe pMOSFETs with high-k gate dielectrics and a TiN gate. A poly-Si/SiO$_2$/Si device is included for comparison.

noisy than the devices with only 0.5-nm interfacial Al$_2$O$_3$. Therefore, it cannot be concluded that Al$_2$O$_3$ performs worse than the other materials. The reason why the Al$_2$O$_3$ contains the highest oxide trap density in this case could be attributed to process induced gate edge damage, a problem sometimes observed with high-k gate dielectrics.[147] Is should be noted that the drain current noise PSD was observed to be of the $1/f^\gamma$-type with the frequency exponent γ in the range 0.9-1.2 for almost all high-k devices in our study. A γ value close to 1 was always observed in strong inversion, whereas some samples of the Al$_2$O$_3$ and the HfO$_2$ devices showed γ up to 1.2 when biased in the subthreshold regime.

An interesting observation, which is further elaborated below, is that the trap density extracted for HfO$_2$ is almost at the same level for devices processed in different batches, with different interface properties and somewhat different deposited thicknesses. However, nMOSFETs seem more sensitive both to the HfO$_2$ thickness as well as the concentration of Hf atoms in the dielectric stack than pMOSFETs. In the work by Simoen and co-workers, significant differences were observed among devices with 3-nm or 5-nm thick HfO$_2$.[24] Moreover, Srinivasan *et al.* report lower 1/f noise for HfSiON than that for HfO$_2$ in case of nMOSFETs but a negligible influence of Hf concentration in pMOSFETs.[31] Giusi *et al.* studied the LF noise in nMOSFETs and reported similar trap densities of HfSiON and SiO$_2$, which were more than one order of magnitude lower than that for HfO_2.[36] Still, the trap densities in HfSiON are generally in excess of 10^{18} $cm^{-3}eV^{-1}$, which is about an order of magnitude higher than that for SiON.[26,31]

7.2.3 Influence of interfacial oxide layer

A thin interfacial layer of SiO_2 is in many cases unavoidably and unintentionally formed between the high-k layer and the Si substrate. Obviously, the presence of an interfacial oxide increases the equivalent oxide thickness and its thickness should therefore be minimized. However, an interfacial SiO_2 layer can also be desirable for improved mobility and interfacial quality. The thickness of the interfacial SiO_2 layer between the high-k and the substrate has been found to be important also for the LF noise properties.[133,137] A large separation of the traps and defects in the high-k layer from the carriers in the channel reduces the $1/f$ noise as fewer carriers can tunnel the long distance. The Coulomb interaction between the charged trap and the channel carriers is also weaker resulting in lower correlated mobility fluctuation noise. But even if a thick SiO_2 interfacial layer (~ 4 nm) is used, deposition of a high-k layer on top of it results in higher $1/f$ noise.[12,13] This suggests that defects propagate from the high-k layer towards the bottom interface, as an exchange of carriers at 4-nm distance is highly unlikely. Min *et al.* have performed a comprehensive investigation of the dependence of the interfacial layer.[148] They found an inverse relationship between oxide trap density and interfacial oxide thickness when the interfacial oxide thickness ranged between 0.8 to 1.8 nm. However, a discontinuity in the trap density at the interface between the high-k layer and the interfacial layer is expected to cause a discontinuity in the $1/f^\gamma$ slope in the noise spectrum according to the McWorther model. Such behaviour was, however, not observed by Min and co-workers.

When the interfacial layer is below 1.0 nm thick, its influence on the $1/f$ noise has been found to be minor according to some reports.[37,43] On the other hand, Crupi *et al.* showed that the mobility fluctuation noise was enhanced when the interfacial layer was thinned from 0.8 nm to 0.4 nm.[39] The reason could for example be a stronger impact of remote phonon scattering as will be discussed later. We have studied the influence of the cleaning prior to ALD of the high-k layer.[43] The devices used a surface $Si_{0.7}Ge_{0.3}$ channel and a gate stack consisting of p^+ poly-SiGe as gate material and $Al_2O_3/HfO_2/Al_2O_3$ as gate dielectrics. Prior to ALD, the $Si_{0.7}Ge_{0.3}$ surface was treated either with an HF-clean or with an HF-clean followed by water-rinse. This was found to cause a difference in the interfacial oxide thickness of around 0.6 nm with a larger thickness (~0.8 nm) for the water rinsed surface. Moreover, the nominal thickness of the bottom Al_2O_3 layer was varied (0.5 or 2 nm).

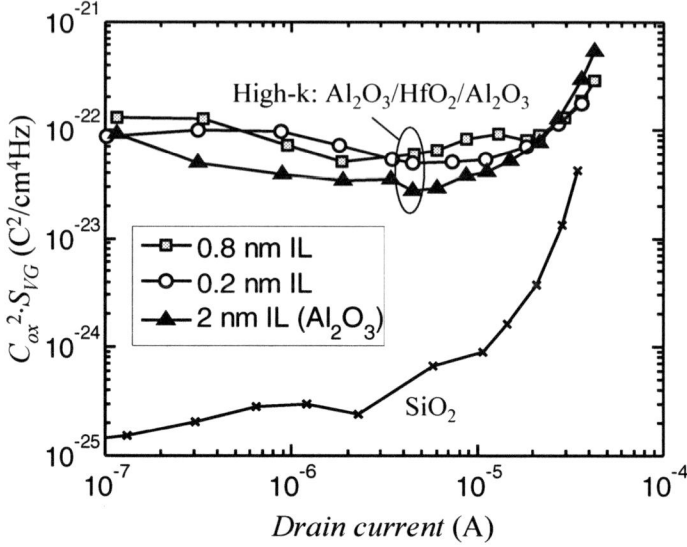

Figure 4-26. Input gate voltage noise plotted for three devices with different thickness of the interfacial layer (IL). The interfacial oxide is composed by Al_2O_3 and SiO_2. A device with SiO_2 gate dielectrics is also shown for comparison.

Comparing the $1/f$ noise characteristics in Fig. 4-26, no significant difference is found between the devices with 0.2 or 0.8 nm interfacial layer (IL). The graph also demonstrates that the high-k devices show significantly increased $1/f$ noise compared to the SiO_2 reference device. The interval of interface layer thicknesses ranges from a few Å up to 1 nm in this study, which may be too narrow in order to observe significant differences in contrast to the report by Crupi *et al.* In any case, the $1/f$ noise was not sensitive to the cleaning prior to the ALD process step in this study. The analysis reveals that the device with 2-nm thick Al_2O_3 at the bottom interface shows a factor of two lower $1/f$ noise than that for the other devices (0.5-nm nominal thick Al_2O_3) at low currents. This may be due to the fact that the traps in the HfO_2 layer are located too far from the channel to contribute to the $1/f$ noise, except at frequencies below 10 Hz, in case the bottom Al_2O_3 layer is 2-nm thick. In a similar type of study, Devireddy *et al.* reported that a stress-relieved preoxide (SRPO) treatment was found to yield lower $1/f$ noise at high gate voltage overdrives than that for devices cleaned by an RCA process before high-k deposition.[37]

7.2.4 Influence of different annealing treatments

An important challenge is to improve the quality of the high-k gate dielectrics. The density of traps and charges in the bulk of the high-k films and at their interfaces needs to be reduced. The introduction of hydrogen annealing, now a standard process step in Si processing, resulted in lower density of interface states at the SiO_2/Si interface and reduced $1/f$ noise in bipolar and MOS transistors.[29,149] The available reports about the impact of annealing on the LF noise characteristics in MOSFETs with high-k gate dielectrics are scarce so far. Srinivasan and co-workers performed a post-deposition anneal (PDA) in a N_2 or NH_3 ambient after deposition of HfO_2 gate dielectrics.[150] The N_2 anneal did not have any significant impact on the LF noise performance, whereas the NH_3 anneal led to one order of magnitude higher LF noise in the nMOSFETs. The pMOSFETs, on the other hand, were less affected but the un-annealed devices showed slightly lower LF noise. Even though annealing in nitrogen can have a positive influence on the high-k gate dielectrics, for example lower EOT values and reduced dopant penetration, incorporation of nitrogen often leads to higher LF noise. Nitrogen is believed to cause defects in the oxide, increasing the number fluctuation noise.

In this work, a novel type of post-metallization annealing was investigated.[151] A forming gas anneal, 10% H_2/90% N_2 at 400 °C, was employed in all high-k devices in this work and is not given particular attention here. Instead, in order to reduce the fixed charge and trap densities in the high-k material (even further), thereby possibly reducing the $1/f$ noise, a novel post-processing step in form of low-temperature water vapour annealing was performed. The water vapour annealing was found to be effective in reducing the negative charge in the Al_2O_3. Fig. 4-27(a) displays the shift in threshold voltage with annealing time and Fig. 4-27(b) demonstrates the effects on the hole mobility. As expected, the mobility increased at low effective field as the oxide charge decreased due to reduced Coulomb scattering. The mobility was found to be maximized after 90-min H_2O annealing. As the annealing continued after that, positive charge was added in the gate dielectrics causing decreased mobility as evidenced in Fig. 4-27(b).

The effects on the $1/f$ noise characteristics were shown to be twofold. The H_2O annealing was not found to reduce the $1/f$ noise itself (at least not when the devices were annealed for 210 min), but the combination of H_2O annealing and a subsequent bake in Argon resulted in improved noise performance, as seen in Fig. 4-28. The slope of the S_I/I^2 curve was found to change with annealing. The curve for the un-annealed device followed a $1/I^\beta$

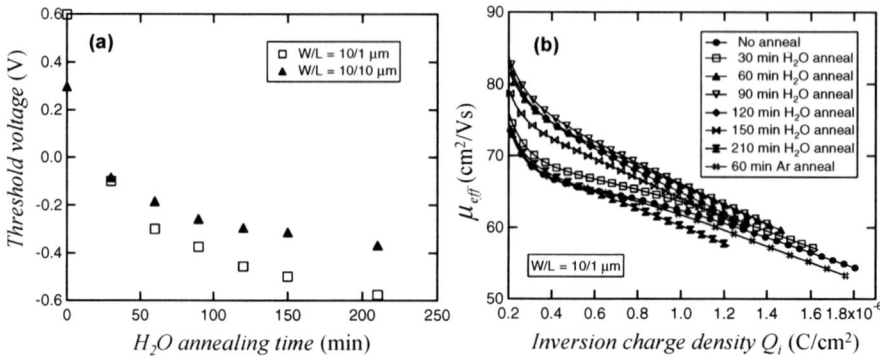

Figure 4-27. (a) Threshold voltage vs. annealing time. The H_2O annealing was carried out at 300 °C. (b) Hole mobility extracted from I_D-V_{GS} curves for various H_2O annealing times. The mobility for the devices annealed in Ar for 60 min after the 210 min H_2O annealing is very similar to that for the un-annealed devices. Reprinted from von Haartman et al.[151] with permission from Elsevier.

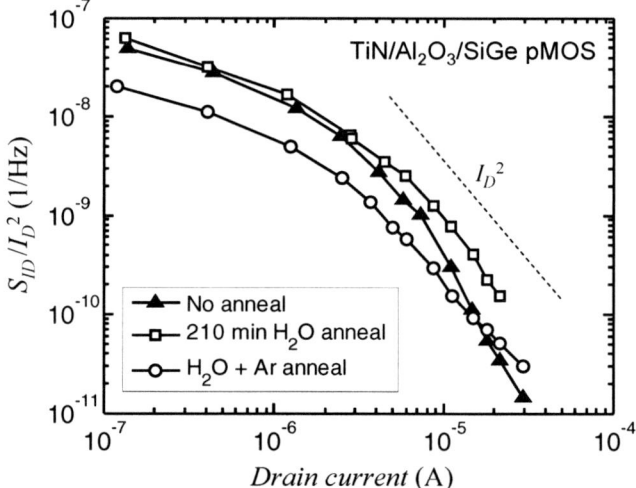

Figure 4-28. Normalized drain current noise at 10 Hz for an un-annealed, a 210 min H_2O annealed and an Ar-treated $Si_{0.8}Ge_{0.2}$ pMOSFET with Al_2O_3 gate dielectrics. $V_{DS} = -50$ mV, $W/L = 10/1$ μm. Reprinted from von Haartman et al.[151] with permission from Elsevier.

behaviour in strong inversion with β around 2.8, whereas β was extracted to be between 1.8 and 2.2 for the H_2O annealed and Ar-treated devices. This difference was attributed to the influence of correlated mobility fluctuations. A negative correlation was found for the un-annealed device, which leads to a steeper decrease of S_I/I^2 with drain current. The negative correlation is consistent with the fact that the gate dielectrics contained a negative charge. The negative charge is reduced upon trapping a hole, which results in a decrease of the Coulomb scattering and consequently a negative correlation between the fluctuating inversion charge density and the fluctuating mobility. The 210 min annealed device, on the other hand, contains a positive charge. Then the mobility reduces upon trapping a hole, and the inversion charge and mobility fluctuations correlate positively. The scattering parameter α was studied versus the threshold voltage of the devices ($\Delta V_T = -\Delta Q_{ox}/C_{ox}$) which was illustrated in Fig. 3-10(b). The maximum magnitude of α was found to be around 1×10^4 Vs/C. The magnitude and sign in front of α depend on the type of traps (acceptor or donor), see Table 3-1 in chapter 3.3.3, as well as the nature of the charge in the gate dielectrics.

These two examples demonstrate that annealing is important for the LF noise characteristics. An optimization of the annealing, deposition, interfacial cleaning and interfacial oxide layer thickness is expected to eventually bring down the 1/f noise of the high-k MOSFETs to acceptable levels. Already today, N_t values in the range 10^{17} cm^{-3} eV^{-1} have been reported, which fulfil the ITRS requirements.

7.2.5 Channel type and material

Most MOSFETs with high-k gate dielectrics are fabricated on a Si substrate. However, other materials, for example Ge, SiGe, GaAs etc, have higher carrier mobilities than Si. One important motivation to use Si is the superior quality of the SiO_2 gate oxide that is formed at the Si surface by oxidation. If the SiO_2 anyway must be replaced, why not replace the channel material also? In the long term, different flavors of CMOS technology in terms of channel materials in combination with high-k are expected to emerge. The purpose here is to discuss the impact of the channel type and material on the LF noise properties. So far, most work has been focused on integration of SiGe with high-k, however some recent results on Ge-on-insulator MOSFETs with HfO_2 are also available. In the latter devices, the trap density was found to be in the range 1×10^{20} cm^{-3}eV^{-1}, roughly one decade higher than that in the Si reference devices.[84] The Ge/interfacial layer was found to of poor quality and considered to be a major reason for the higher noise in these devices. For the former type of device with a SiGe channel, the technology is more mature. The device technology group at KTH has made important efforts in fabricating and evaluating such devices; the results of this work will be presented here.

High-k MOSFETs utilizing a surface SiGe channel have been investigated. A compressively strained SiGe channel is desired for its superior hole mobility compared to Si. Since no oxidation step is performed in the MOSFETs with high-k gate dielectrics, the Si-cap might no longer be necessary for maintaining a low interface state density. A surface channel is advantageous since the parasitic current limiting the drive current enhancement in a buried SiGe channel transistor is eliminated. However, the interface state density is almost one order of magnitude higher in the fabricated surface SiGe transistors, likely due to the formation of an interfacial oxide layer; see Table 4-3. Still, the $1/f$ noise performance is not deteriorated in the SiGe channel transistors. Comparing the two $HfAlO_x$ transistors in Fig. 4-25, no significant difference in noise level is found. In fact, by studying the drain current noise versus gate voltage overdrive, a noise reduction by a factor around two is observed for the SiGe transistors. This is demonstrated in Fig. 4-29 showing the Hooge parameter plotted as a function of gate voltage overdrive. Note that the devices with poly-SiGe gate show higher noise compared to the devices with TiN gate, which will be further addressed in the next section.

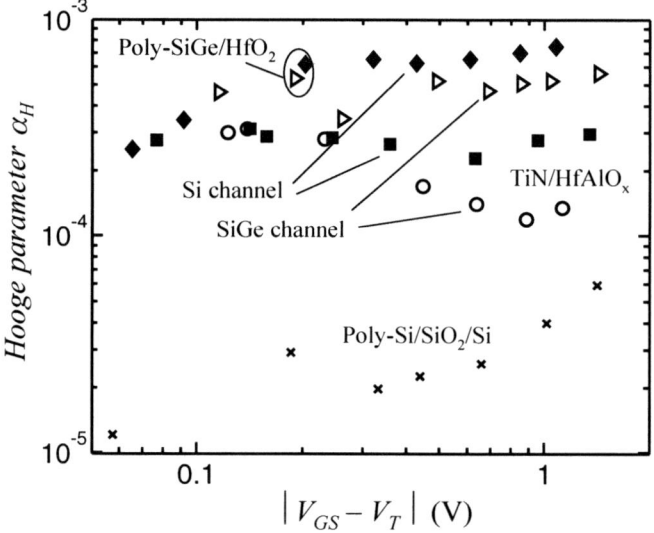

Figure 4-29. Hooge parameter vs. gate voltage overdrive for high-k pMOSFETs with a SiGe or Si channel. A poly-Si/SiO$_2$/Si reference device is shown for comparison.

An important relationship is found if the Hooge parameter is studied versus low-field mobility as shown Fig. 4-30. The graph shows the Hooge parameter, extracted at $V_{GT} = 1$ V, for all high-k MOSFETs investigated in this work. As seen, lower α_H values are obtained for the SiGe devices than that for the Si devices with the same gate stack. The dispersion in

4. 1/f noise performance of advanced CMOS devices

α_H and μ_0 values is not dramatic, even though some of the extracted values are from transistors with different gate lengths. An interesting observation in Fig. 4-30 is that low noise is correlated to a high mobility and vice versa. This indicates that the $1/f$ noise mechanism is related to some scattering mechanism. A more detailed analysis carried out elsewhere demonstrates that the phonon scattering can explain the mobility differences among the devices with different gate stack and channel material, which suggests that mobility fluctuation noise prevail.[41] High-k transistors suffer from an additional phonon scattering contribution, the remote phonons in the high-k gate dielectrics, which might be one explanation why these devices show higher $1/f$ noise than conventional CMOS transistors.

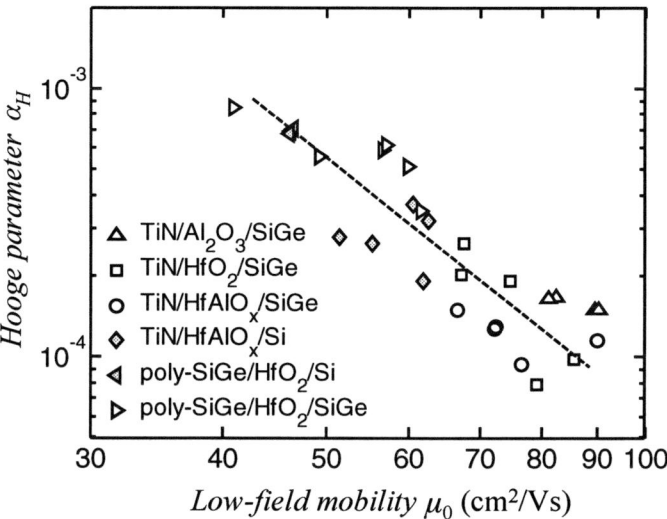

Figure 4-30. Hooge parameter values extracted at $V_{GS} - V_T = -1$ V studied vs. low-field mobility for various pMOSFETs with high-k gate dielectrics.

Note that Simoen *et al.* also found a similar relationship when the trap density was plotted versus the mobility.[24] In their study, they compared the parameters for different samples of the same device type and found a correlation between low trap density and high mobility. In Fig. 4-30, the parameters are compared for several samples of different devices. The channel, gate dielectric and gate electrode materials differ in the studied devices. The general trend is that a low α_H is correlated to a high μ_0 also when a certain device type is studied. Thus, a random density of defects and imperfections cause correlated variations in both these parameters. However, the relationship is weak or different in some cases. Two important conclusions can be drawn from this study. First, the mobility and the $1/f$

noise are affected by the same mechanism. Secondly, optimization of the mobility likely results in lower $1/f$ noise, which of course is highly desirable. This can be expected to hold for a family of devices with similar technology and similar densities of defects.

The high $1/f$ noise level in the high-k transistors is a problem, which could disqualify them to be used in future analog circuits according to Fig. 4-3. Techniques to lower the $1/f$ noise should therefore be sought out. Buried SiGe channel pMOSFETs have been successful in lowering the $1/f$ noise in the past. Studies performed on high-k gate dielectrics pMOSFETs indicate that the $1/f$ noise is reduced when a forward substrate bias is applied, as shown in Fig. 4-31, which suggest that a buried channel can be beneficial also for these types of devices. In the first attempt to fabricate buried SiGe channel transistors with HfO_2 gate dielectrics, no significant reduction in the noise level was observed.[45] According to Ghibaudo and Chroboczek,[81] $1/f$ noise originating from trapping/release phenomena in the gate dielectrics is not necessarily reduced although most of the current flows in a buried SiGe channel. If the trap densities in the gate dielectrics are similar one could also expect similar noise levels. However, neither the quality of the compressively strained SiGe layer nor the high-k was optimized. Further studies should therefore be undertaken.

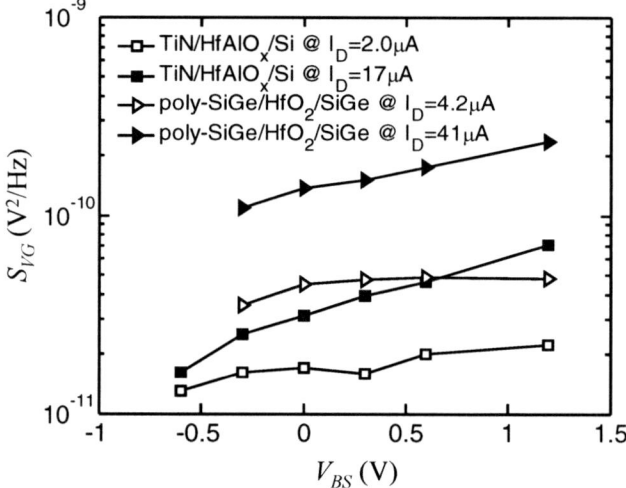

Figure 4-31. Input gate voltage noise at different I_D plotted vs. body-source voltage V_{BS} for two different pMOSFETs with high-k gate dielectrics. A forward substrate bias reduces the $1/f$ noise in the studied devices, especially at higher drain currents. $W/L = 10/1$ μm, $V_{DS} = -50$ mV, $f = 10$ Hz.

7.2.6 1/f noise modeling

So far, most evidence has pointed to a number fluctuation noise origin in MOSFETs with high-k gate dielectrics. But as the trap densities have decreased to more moderate levels, thus generating lower number fluctuation noise, the mobility fluctuation noise have been found to become dominant in many of the studied pMOSFETs.[31,39,41] In this work, the devices with poly-SiGe gate or TiN gate and HfAlO$_x$ gate dielectrics are well modeled with Hooge's model. An example was given in Fig. 4-29 showing that α_H is relatively constant in accordance with Hooge's model. However, the normalized drain current noise in the devices with TiN gate and HfO$_2$ or Al$_2$O$_3$ gate dielectrics showed a stronger I_D dependence (see Fig. 4-25), the 1/f noise is then better explained with the number fluctuation noise model. This indicates that the trap density is higher in those gate dielectric structures.

An important difference regarding the noise mechanisms and modeling compared to standard transistors with SiO$_2$ gate dielectrics concerns the correlated mobility fluctuations. In Fig. 4-32, the gate voltage noise is plotted versus I_D/g_m for TiN gated transistors with high-k gate dielectrics. The solid lines are simulations using the model in Eq. (3-30) with the following α values: 4×10^4 (HfAlO$_x$/Si), 1×10^4 (HfAlO$_x$/SiGe), -8×10^3 (Al$_2$O$_3$/SiGe), -1×10^3 or 6×10^3 Vs/C (HfO$_2$/SiGe). Negative values of the scattering parameter are found which is unusual for standard CMOS transistors. This indicates for example that the dominant type of trap (acceptor or donor) differs for different gate dielectric materials.

As seen, the HfO$_2$ and Al$_2$O$_3$ devices are difficult to model over the whole studied bias range due to the U-shaped gate voltage noise curve, which suggest that two different noise mechanisms are involved. In the subthreshold region and at low gate voltage overdrives, the number fluctuation noise dominates. Thus, the density of traps in the high-k gate dielectrics is important in this region. As shown in Table 4-3, the oxide charge density is higher in the HfO$_2$ and Al$_2$O$_3$ gate stacks, which points to a higher trap density. In the strong inversion region, on the other hand, the gate voltage noise curves for all the devices studied in Fig. 4-32 approach each other, irrespective of the noise level in weak inversion. This suggests that another noise mechanism is dominant in this region of operation. Mobility fluctuations take over as the dominant noise source in strong inversion. The mobility fluctuations are less sensitive to the gate dielectric material, but instead more dependent on the channel material and gate material since these two components have a large impact on the mobility.

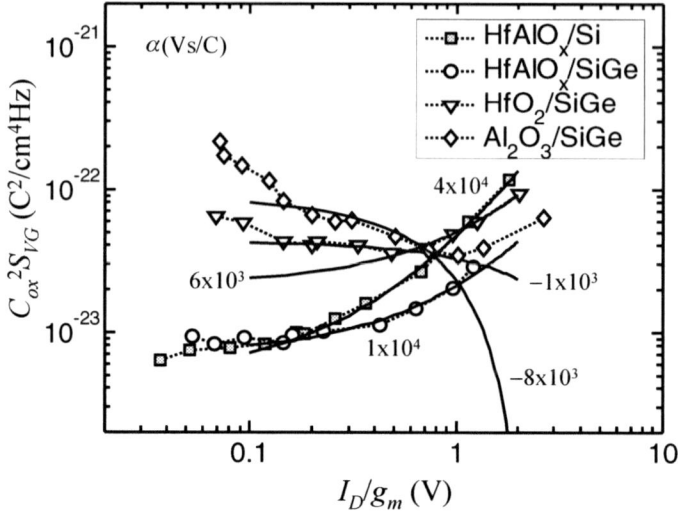

Figure 4-32. Input gate voltage noise plotted vs. I_D/g_m. The solid lines are simulations using Eq. (3-30). The α values used in the model are indicated. Reprinted from von Haartman et al.[41] with permission from IEEE.

7.2.7 Summary and future outlook

MOSFETs with high-k gate dielectrics show increased $1/f$ noise compared to transistors using SiO_2 due to large defect densities originating from the high-k materials and possibly also due to mobility fluctuations originating from the remote phonon scattering. In terms of high-k materials, $HfAlO_x$ and HfSiON often contain lower trap densities N_t than HfO_2 and Al_2O_3 do. HfO_2 is the main contender to replace SiO_2 and has therefore been studied more extensively and has been further optimized compared to the other materials. As with technology shifts, some time must be expended to learn the new materials and processing methods. We can expect that the $1/f$ noise will be reduced as the technology becomes more mature. Today, the $1/f$ noise of HfO_2 gate dielectric devices has reached acceptable levels according to some reports (N_t in the range of 4×10^{17} $cm^{-3}eV^{-1}$), thus indicating that the ITRS requirements can be met.[37] An important finding, which we will discuss in more detail in the next section, is that the reduced $1/f$ noise found in these studies partly is a consequence from replacing the poly-Si with a metal gate material. Concepts to reduce the $1/f$ noise, such as using a buried channel, also need to be explored more in high-k devices. In this context, it is encouraging with the results presented in this work where reduced device $1/f$ noise was found when the substrate was forward biased. In conclusion, based on our results the noise properties of high-k gate dielectrics seem promising for future generations of MOSFETs, yet some problems remains to be solved.

8. METAL GATE DEVICES

8.1 Metal gate materials and characteristics

The use of a metal gate instead of the poly-Si eliminates the problem with dopant penetration through the gate dielectric, Fermi-level pinning that can raise the threshold voltage of transistors with a poly-Si/high-k gate stack, and poly-depletion that reduces the effective oxide capacitance.[14,152,153] The sheet resistance, which is important for the high-frequency properties of the device, can potentially be lower in a metal gate technology. Furthermore, if a mid-gap metal gate is used in combination with high-k gate dielectrics, the mobility degradation due to remote phonons has been reported to diminish due to more effective screening of the soft phonon modes from coupling to the channel.[154] However, the work function of the metal gate electrode material must be appropriate to give the correct threshold voltage; thermal stability and process integration are other issues. Several metal gate candidates have been investigated such as TaN, TaSiN, Mo, Ru, TiAlN (see Sjöblom et al.[153] and references therein), TiN,[14,142] and fully silicided (FUSI) gates using for example NiSi.[152] We have studied TiN in combination with high-k in this work. TiN was found to be attractive both due to a higher hole mobility in the devices in comparison with the poly-SiGe gated ones, as shown in Fig. 4-23, and lower $1/f$ noise.

8.2 $1/f$ noise characteristics

Replacing the poly-Si with a metal or silicide such as TiN, TaSiN, NiSi or TaN is highly desired and is currently a vivid research topic. As the gate dielectrics becomes thinner and thinner, the impact of the gate material on the device properties is expected to increase. For high-k MOSFETs, there are several reports indicating lower $1/f$ noise when a metal gate is used instead of a poly-Si gate. It was shown in the previous section that the TiN gate was found to be favourable in comparison with the poly-SiGe gate in terms of both mobility and $1/f$ noise performance in the transistors with high-k gate dielectrics. Recent reports confirm these exciting results for PVD TaN, NiSi,[155] and ALD TaSiN.[37] The lowering of the $1/f$ noise is mainly observed in the strong inversion regime. Fig. 4-33 illustrates the normalized drain current noise plotted versus gate voltage overdrive for poly-SiGe and TiN gated transistors with $Al_2O_3/HfO_2/Al_2O_3$ stacks as gate dielectrics. As observed, the normalized drain current noise is at the same level for all the HfO$_2$ gate dielectric pMOSFETs at low bias, whereas the TiN gated device shows significantly reduced noise in strong inversion.

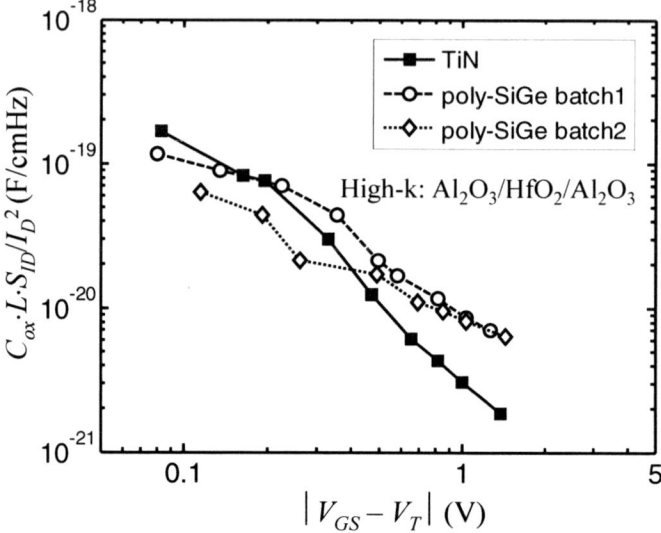

Figure 4-33. Normalized drain current noise studied vs. gate voltage overdrive for pMOSFETs with high-k gate dielectrics in form of Al$_2$O$_3$/HfO$_2$/Al$_2$O$_3$ stacks and using TiN (metal gate) or poly-SiGe as gate electrode material.

The 1/*f* noise lowering has been reported for TiN, TaSiN, TaN and NiSi gate materials, which might suggest that the noise reduction is due to an inherent property of the metal gate. The metal gate is known to alleviate the effect of remote phonon scattering on the hole mobility. The same mechanism is here proposed to explain the 1/*f* noise reduction. The mobility fluctuation noise is assumed to be mainly generated in the phonon scattering. Therefore, a high-k device with a metal gate can be less noisy than the same device with a poly-Si gate. On the other hand, the traps and charges at the gate-dielectric interface might also be better screened by a metal gate, as indicated by Srinivasan *et al.*[155]

Another explanation of the lower 1/*f* noise is that the metal gate has some positive impact on the oxide trap density. The difference in metal gate deposition compared to a poly-Si process could affect the quality of the gate electrode/gate dielectric interface as well as the underlying gate oxide as suggested by Devireddy *et al.*[37] Srinivasan *et al.* instead speculate in a recent publication that the noise reduction is attributed to a lower concentration of oxygen-vacancy related defects when a metal gate is used.[38] However, an unoptimized deposition process can induce additional traps in the gate dielectrics. Lee et al. investigated pMOSFETs on SOI with a Mo gate.[22] The Mo film was deposited using DC magnetron sputtering and a 2.5-nm thick gate oxide was used. The results pointed to a relatively high noise level ($N_t \sim 1 \times 10^{18}$ cm^{-3}eV^{-1}), roughly one order of magnitude higher than that of conventional poly-Si gated MOSFETs.

In summary, further experiments are needed to pinpoint the exact origin of the lower $1/f$ noise in metal gate MOSFETs; the results on using metal gates in combination with high-k gate dielectrics are very promising so far.

9. MULTIPLE GATE DEVICES

9.1 Device structures and characteristics

The multiple-gate MOSFET concept is a very attractive solution for ultra-scaled CMOS technologies at the 45 nm node and beyond. Several multiple-gate architectures have been proposed, such as back gated ultra-thin body SOI,[104,116,156] gate-all-around MOSFETs,[157] FinFETs,[158] and Omega FETs.[111] The advantages by using multiple gates are mainly improved drive current, better control of the short channel effect and reduced drain-induced-barrier-lowering. By increasing the number of gates, their electrostatic control of the channel is improved correspondingly. In chapter 6.1.2 about FD SOI devices, we mentioned that a silicon body thickness corresponding to 1/3 of the gate length is required to control the short-channel effect. For a double gate device, like the FinFET, the required body thickness t_{Si} is $2L/3$, increasing to $t_{Si} \approx L$ for a tri-gate device (for example Omega FETs). Thus, shorter gate lengths can be used for a certain body thickness if the number of gates is increased.

FinFETs have received a lot of attention because of their excellent performance and the relatively simple fabrication. Other multiple-gate devices like the Omega FET use a very similar fabrication process. The idea is to make a thin fin or nano-wire in the Si body of a SOI substrate. The gate stack is deposited on top and patterned. The conducting channels are formed at the sidewall of the silicon fin and on top of the fin (for a tri-gate device) as illustrated in Fig. 4-34. The sidewall channels are in the (110) crystal direction, which is different from a planar CMOS device that normally uses the (100) direction. The (110) crystal direction is beneficial for hole transport providing a significant improvement of around 100% for the hole mobility.[158] The electron mobility, on the other hand, is lower in the (110) crystal direction.

When the fin thickness is below ~10 nm, volume inversion effects start to play a role in confining the inversion carriers to the interior of the body instead of at the surfaces. By inverting the interior of the body, the scattering and the noise generation related to the gate dielectric interfaces are avoided. Higher mobilities have been reported in such devices[104,116,159] and there are indications that the 1/f noise can be lower as well.[113]

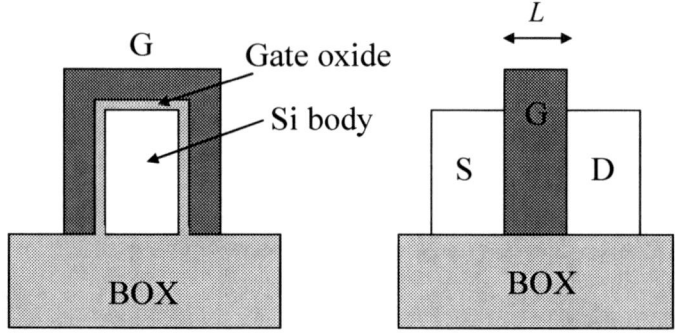

Figure 4-34. Schematic cross section (left) and side view (right) of a tri-gate MOSFET.

9.2   1/f noise characteristics

Presently, there are only a few publications available about the LF noise properties of multiple gate MOSFETs leaving many issues to be explored in the future. The FinFET conducts current in a vertical mode, which may lead to degraded 1/f noise performance according to recent reported results.[160,161] The etching of the sidewalls of the fin could lead to higher micro-roughness and increased trap densities at the gate oxide/channel interface, which might be one explanation for the higher 1/f noise. Also, the gate area of FinFETs is very small, thus the devices might suffer from edge effects. On the other hand, by using a thin Si body (< ~10nm), volume inversion effects could potentially result in (much) lower 1/f noise. Significantly lower noise was reported for gate-all-around MOSFETs under some bias conditions,[157] although the film thickness was too large in that case (100 nm) to produce volume inversion. In the few reports about the low-frequency noise characteristics of FinFETs so far,[29,34,162] no evidence of lower noise due to volume inversion has been observed. As seen in Fig. 4-3, low N_t values are found for some FinFETs that were annealed in hydrogen, but the values are not lower than that in bulk CMOS. Therefore, further development of the multiple-gate MOS-technology is necessary, and further LF noise investigations of multiple gate MOSFETs are urgently asked for.

Recent results on a special four-gate FET, which combines the JFET and MOSFET operation modes, indicate a 1/f noise reduction by one order of magnitude for volume conduction compared to surface conduction.[163]

4. 1/f noise performance of advanced CMOS devices 161

Akarvardar and co-workers discussed a transition from number fluctuation noise to mobility fluctuation noise when the conduction channel moved from the surface towards the center of the Si film. However, g-r noise was observed in both conduction regimes and the frequency exponent was generally smaller ($\gamma \approx 0.7$ compared to $\gamma \approx 1$) for the volume conduction mode. This indicates that the total noise likely is a combination of number and mobility fluctuations for both conduction modes. In fact, the mobility fluctuation noise is reduced in the volume conduction mode as well. The value of the Hooge parameter was found to be 2×10^{-5}, which is a typical value for a Si MOSFET with pure SiO_2 as gate oxide. Nevertheless, the results obtained by Akarvardar *et al.* confirm the potential with volume conduction for minimization of the 1/*f* noise.

SUMMARY

This chapter has presented an overview of advanced CMOS concepts including ultra-scaled MOSFETs, buried/surface SiGe channels, strained-Si, high-k gate dielectrics, nitrided SiO_2, SOI, multiple-gate architectures, metal gates, Ge-on-insulator, and Schottky Barrier MOSFETs. We have in detail addressed the low-frequency noise properties in each of these CMOS technologies. The most important findings are summarized below.

Technology solutions to reduce the LF noise:
- Buried channel for the drain current. Examples: buried channel MOSFETs fabricated by counter doping the surface of the substrate, buried SiGe channel pMOSFETs, JFETs, AM SOI MOSFETs, and (potentially) multiple-gate devices showing volume inversion. Moreover, the channel carriers will be located further away from the gate oxide interface in case of a forward bias on the substrate (or an appropriate back gate bias for a SOI MOSFET), which is expected to lower the LF noise.
- Metal gates can lower the 1/*f* noise in MOSFETs with high-k gate dielectrics.
- Improved quality of the gate dielectrics, which can be achieved by annealing or using improved growth/deposition and cleaning methods.
- Larger device area and thinner gate dielectrics. However, thinning the gate dielectrics enhances the gate leakage current and the noise associated with it.
- Strained-Si has been reported to show reduced 1/*f* noise.
- Lower S/D resistance (important in the high-current regime).

Technology solutions that could or will lead to higher LF noise:
- High-k gate dielectrics.
- Nitrided SiO_2.
- SOI, due to floating body effects (mainly PD devices) and front-back gate coupling effects (FD).
- Strained-Si (locally stressed).
- Strained-Si devices fabricated on SiGe virtual substrates, due to dislocations and Ge-outdiffusion.
- Ge channel MOSFETs.
- Schottky Barrier MOSFETs.
- Downscaling of the MOSFET gate width and gate length.
- Complex device structures.
- Technology that leads to high defect densities (especially in the gate oxide and the channel) and poor quality of the gate oxide/channel interface.

Technology solutions that improve the performance of MOSFETs without sacrificing the noise performance (and not in the first group):
- Changing channel orientation.
- Strained-Si (could in same cases lead to either lower or higher LF noise).
- S/D engineering (as long as the LF noise generated in the channel dominates).

REFERENCES

1. R. H. Dennard, F. H. Gaensslen, H. N. Yu, V. L. Rideout, E. Bassous, and A. LeBlanc, Ion implanted MOSFETs with very short channel lengths, in *IEDM Tech. Dig.*, 1973, pp. 152-155.
2. R. H. Dennard, F. H. Gaensslen, H.-N. Yu, V. L. Rideout, E. Bassous, and A. R. LeBlanc, Design of ion-implanted MOSFET's with very small physical dimensions, *IEEE J. Solid-State Circuits*, **SC-9**, 256-268 (1974).
3. L. D. Yau, A simple theory to predict the threshold voltage of short-channel IGFETs, *Solid-State Electron.* **17**, 1059-1063 (1974).
4. G. Baccarani, M. R. Wordeman, and R. H. Dennard, Generalized scaling theory and its application to a 1/4 micrometer MOSFET design, *IEEE Trans. Electron Devices* **ED-31**, 452-462 (1984).
5. International Technology Roadmap for Semiconductors (ITRS), 2005 update, http://www.itrs.net
6. A. A. Abidi, RF CMOS comes of age, in *Proc. Symp. VLSI Circuits*, 2003, pp. 113-116.
7. A. A. Balandin, *Noise and fluctuations control in electronic devices* (American Scientific Publishers, Stevenson Ranch, CA, 2002).
8. C. Claeys and E. Simoen, Impact of advanced processing modules on the low-frequency noise performance of deep-submicron CMOS technologies, *Microelectron. Reliab.* **40**, 1815-1821 (2000).

9. S. Okhonin, M. A. Py, B. Georgescu, H. Fischer, and L. Risch, DC and low-frequency Noise Characteristics of SiGe P-Channel FET's Designed for 0.13-µm Technology, *IEEE Trans. Electron Devices* **46**, 1514-1517 (1999).
10. S. J. Mathew, G. Niu, W. B. Dubbelday, and J. D. Cressler, Characterization and Profile Optimization of SiGe pFET's on Silicon-on-Sapphire, *IEEE Trans. Electron Devices* **46**, 2323-2332 (1999).
11. M. J. Prest, M. J. Palmer, G. Braithwaite, T. J. Grasby, P. J. Phillips, O. A. Mironov, E. H. C. Parker, and T. E. Whall, Si/Si$_{0.64}$Ge$_{0.36}$/Si pMOSFETs with enhanced voltage gain and low 1/f noise, in *Proc. ESSDERC*, 2001, pp. 179-182.
12. T. Horikawa, N. Yasuda, W. Mizubayashi, K. Iwamoto, K. Tominaga, K. Akiyama, K. Yamamoto, H. Hisamatsu, H. Ota, T. Nabatame, and A. Toriumi, Low frequency noise characteristics in HfAlO$_x$/SiO$_2$ n-MOSFETs, in *Proc. Vol. 1 Electrochem Soc. Meeting*, 2004, pp. 292-303.
13. B. Min, S. P. Devireddy, Z. Çelik-Butler, F. Wang, A. Zlotnicka, H.-H. Tseng, and P. J. Tobin, Low-frequency noise in submicrometer MOSFETs with HfO$_2$, HfO$_2$/Al$_2$O$_3$ and HfAlO$_x$ Gate Stacks, *IEEE Trans. Electron Devices* **51**, 1679-1687 (2004).
14. B. Guillaumot, X. Garros, F. Lime, K. Oshima, B. Tavel, J. A. Chroboczek, P. Masson, R. Truche, A. M. Papon, F. Martin, J. F. Damlencourt, S. Maitrejean, M. Rivoire, C. Leroux, S. Christoloveanu, G. Ghibaudo, J. L. Autran, T. Skotnicki, and S. Deleonibus, 75nm damascene metal gate and high-k integration for advanced CMOS devices, in *IEDM Tech. Dig.*, 2002, pp. 355-358.
15. J. Chang, A. A. Abidi, and C. R. Viswanathan, Flicker noise in CMOS transistors from subthreshold to strong inversion at various temperatures, *IEEE Trans. Electron Devices*, **41**, 1965-1971 (1994).
16. M. Fadlallah, G. Ghibaudo, J. Jomaah, M. Zoaeter, and G. Guégan, Static and low frequency noise characterization of surface- and buried-mode 0.1 µm P and NMOSFETS, *Microelectron. Reliab.* **42**, 41-46 (2002).
17. X. Li, C. Barros, E. P. Vandamme, and L. K. J. Vandamme, Parameter extraction and 1/f noise in a surface and a bulk-type, p-channel LDD MOSFET, *Solid-State Electron.* **37**, 1853-1862 (1994).
18. M. Marin, M. J. Deen, M. de Murcia, P. Llinares, and J. C. Vildeuil, Effects of body biasing on the low frequency noise of MOSFETs from 130 nm CMOS technology, *IEE Proc.-Circuits Devices Syst.* **151**, 95-101 (2004).
19. A. Mercha, E. Simoen, and C. Claeys, Impact of high vertical electric field on low-frequency noise in thin-gate oxide MOSFETs, *IEEE Trans. Electron Devices* **50**, 2520-2527 (2003).
20. M. Valenza, A. Hoffmann, D. Sodini, A. Laigle, F. Martinez, and D. Rigaud, Overview of the impact of downscaling technology on 1/f noise in p-MOSFETs to 90nm, *IEE Proc.-Circuits Devices Syst.* **151**, 102-110 (2004).
21. A. K. M. Ahsan and D. K. M. Schroder, Impact of post-oxidation annealing on low-frequency noise, threshold voltage, and subthreshold swing of p-channel MOSFETs, *IEEE Electron Device Lett.* **25**, 211-213 (2004).
22. J.-S. Lee, D. Ha, Y.-K. Choi, T.-J. King, and J. Bokor, Low-frequency noise characteristics of ultrathin body p-MOSFETs with molybdenum gate, *IEEE Electron Device Lett.* **24**, 31-33 (2003).
23. Y. Nemirovsky, I. Brouk, and C. G. Jakobson, 1/f noise in CMOS transistors for analog applications, *IEEE Trans. Electron Devices* **48**, 921-927 (2001).
24. E. Simoen, A. Mercha, C. Claeys, and E. Young, Correlation between the 1/f noise parameters and the effective low-field mobility in HfO$_2$ gate dielectric n-channel metal-oxide-semiconductor field-effect transistors, *Appl. Phys. Lett.* **85**, 1057-1059 (2004).

25. F. Dieudonné, S. Haendler, J. Jomaah, and F. Balestra, Low frequency noise in 0.12 μm partially and fully depleted SOI technology, *Microelectron. Reliab.* **43**, 243-248 (2003).
26. B. Min, S. P. Devireddy, and Z. Çelik-Butler, Low-frequency noise characteristics of HfSiON gate-dielectric metal-oxide-semiconductor-field-effect transistors, *Appl. Phys. Lett.* **86**, 082102 (2005).
27. F. Dieudonné, S. Haendler, J. Jomaah, and F. Balestra, Low frequency noise and hot-carrier reliability in advanced SOI MOSFETs, *Solid-State Electron.* **48**, 985-997 (2004).
28. E. Simoen, A. Mercha, C. Claeys, N. Lukyanchikova, and N. Garbar, Critical discussion of the front-back gate coupling effect on the low-frequency noise in fully depleted SOI MOSFETs, *IEEE Trans. Electron Devices* **51**, 1008-1016 (2004).
29. J.-S. Lee, Y.-K. Choi, D. Ha, S. Balasubramanian, T.-J. King, and J. Bokor, Hydrogen annealing effect on DC and low-frequency noise characteristics in CMOS FinFETs, *IEEE Electron Device Lett.* **24**, 186-188 (2003).
30. Y. Akue Allogo, M. Marin, M. de Murcia, P. Llinares, and D. Cottin, $1/f$ noise in 0.18 μm technology n-MOSFETs from subthreshold to saturation, *Solid-State Electron.* **46**, 977-983 (2002).
31. P. Srinivasan, E. Simoen, L. Pantisano, C. Claeys, and D. Misra, Impact of high-k gate stack material with metal gates on LF noise in n- and p-MOSFETs, *Microelectron. Eng.*, **80**, 226-229 (2005).
32. N. Lukyanchikova, N. Garbar, M. Petrichuk, E. Simoen, and C. Claeys, Flicker noise in deep submicron nMOS transistors, *Solid-State Electron.* **44**, 1239-1245 (2000).
33. E. Simoen, G. Eneman, P. Verheyen, R. Delhougne, R. Loo, K. De Meyer, and C. Claeys, On the beneficial impact of tensile-strained silicon substrates on the low-frequency noise of n-channel metal-oxide-semiconductor transistors, *Appl. Phys. Lett.* **86**, 223509, (2005).
34. V. Subramanian, A. Mercha, A. Dixit, K. G Anil, M. Jurczak, K. De Meyer, S. Decoutere, H. Maes, G. Groeseneken, and W. Sansen, Geometry dependence of $1/f$ noise in N- and P- Channel MuGFETs, in *Proc. Int. Conf. Noise and Fluctuations (ICNF)*, 2005, pp. 279-282.
35. K. W. Chew, K. S. Yeo, and S.-F. Chu, Effect of technology scaling on the $1/f$ noise of deep submicron PMOS transistors, *Solid-State Electron.* **48**, 1101-1109 (2004).
36. G. Giusi, F. Crupi, C. Pace, C. Ciofi, and G. Groeseneken, Comparative study of drain and gate low-frequency noise in nMOSFETs with hafnium-based gate dielectrics, *IEEE Trans. Electron Devices* **53**, 823-828 (2006).
37. S. P. Devireddy, B. Min, Z. Çelik-Butler, H. H. Tseng, P. J. Tobin, F. Wang, and A. Zlotnicka, Low-frequency noise in TaSiN/HfO$_2$ nMOSFETs and the effect of stress-relieved preoxide interfacial layer, *IEEE Trans. Electron Devices* **53**, 538-544 (2006).
38. P. Srinivasan, E. Simoen, R. Singanamalla, H. Y. Yu, C. Claeys, and D. Misra, Gate electrode effect on low-frequency (1/f) noise in p-MOSFETs with high-κ gate dielectrics, *Solid-State Electron.* **50**, 992-998 (2006).
39. F. Crupi, P. Srinivasan, P. Magnone, E. Simoen, C. Pace, D. Misra, and C. Claeys, Impact of the interfacial layer on the low-frequency (1/f) noise behaviour of MOSFETs with advanced gate stacks, *IEEE Electron Device Lett.* **27**, 688-691 (2006).
40. M. von Haartman, A.-C. Lindgren, P.-E. Hellström, B. G. Malm, S.-L. Zhang, and M. Östling, $1/f$ noise in Si and Si$_{0.7}$Ge$_{0.3}$ pMOSFETs, *IEEE Trans. Electron Devices* **50**, 2513-2519 (2003).
41. M. von Haartman, B. G. Malm, and M. Östling, Comprehensive study on low-frequency noise and mobility in Si and SiGe pMOSFETs with high-κ gate dielectrics and TiN gate, *IEEE Trans. Electron Devices* **53**, 836-843 (2006).
42. M. von Haartman, J. Hållstedt, J. Seger, B. G. Malm, P.-E. Hellström and M. Östling, Low-frequency noise in SiGe channel pMOSFETs on ultra-thin body SOI with Ni-

silicided source/drain, in *Proc. 18th Int. Conf. Noise and Fluctuations (ICNF)*, 2005, pp. 307-310.
43. M. von Haartman, D. Wu, B. G. Malm, P.-E. Hellström, S.-L. Zhang and M. Östling, Low-frequency noise in Si$_{0.7}$Ge$_{0.3}$ surface channel pMOSFETs with ALD HfO$_2$/Al$_2$O$_3$ gate dielectrics, *Solid-State Electronics* **48**, 2271-2275 (2004).
44. M. von Haartman, D. Wu, P.-E. Hellström, S.-L. Zhang and M. Östling, Low-frequency noise in Si$_{0.7}$Ge$_{0.3}$ surface channel pMOSFETs with a metal/high-κ gate stack, in *Proc. 17th Int. Conf. Noise and Fluctuations (ICNF)*, 2003, pp. 381-384.
45. M. von Haartman, B. G. Malm, P.-E. Hellström and M. Östling, Noise in Si and SiGe MOSFETs with high-k gate dielectrics, in *Proc. 18th Int. Conf. Noise and Fluctuations (ICNF)*, 2005, pp. 225-230.
46. M. von Haartman, Ph. D. Thesis, KTH, Royal Institute of Technology, Sweden, 2006.
47. G. Ghibaudo, and T. Boutchacha, Electrical noise and RTS fluctuations in advanced CMOS devices, *Microelectron. Reliab.* **42**, 573-582 (2002).
48. R. Brederlow, W. Weber, D. Schmitt-Landsiedel, and R. Thewes, Fluctuations of the low frequency noise of MOS transistors and their modeling in analog and RF-circuits, in *IEDM Tech. Dig.*, 1999, pp. 159-162.
49. M. Sandén, O. Marinov, M. J. Deen, and M. Östling, A new model for the low-frequency noise and the noise level variation in polysilicon emitter BJTs, *IEEE Trans. Electron Devices* **49**, 514-520 (2002).
50. J. Brini, G. Ghibaudo, G. Kamarinos, and O. Roux-dit-Buisson, Scaling down and low-frequency noise in MOSFET's: are the RTS's the ultimate components of the 1/f noise?, *AIP Conf. Proc.* **282**, 31-48 (1993).
51. C. Claeys, A. Mercha, and E. Simoen, Low-frequency noise assessment for deep submicrometer CMOS technology nodes, *J. Electrochem. Soc.* **151**, G307-G318 (2004).
52. L. K. J. Vandamme, X. Li, and D. Rigaud, 1/f noise in MOS devices, mobility or number fluctuations?, *IEEE Trans. Electron Devices* **41**, 1936-1945 (1994).
53. A. van der Ziel, *Noise in solid state devices and circuits* (John Wiley & Sons, New York, 1986).
54. M. Bhat, D. J. Wristers, L.-K. Han, J. Yan, H. J. Fulford, and D.-L. Kwong, Electrical properties and reliability of MOSFET's with rapid thermal NO-nitrided SiO$_2$ gate dielectrics, *IEEE Trans. Electron Devices*, **42**, 907-914 (1995).
55. M. A. Schmidt, F. L. Terry, Jr., B. P. Mathur, and S. D. Senturia, Inversion layer mobility of MOSFET's with nitrided oxide gate dielectrics, *IEEE Trans. Electron Devices*, **35**, 1627-1632 (1988).
56. T. Ishihara, K. Matsuzawa, M. Takayanagi, and S. Takagi, Comprehensive understanding of electron and hole mobility limited by surface roughness scattering in pure oxides and oxynitrides based on correlation function of surface roughness, *Jpn. J. Appl. Phys.* **41**, 2353-2358 (2002).
57. P. Morfouli, G. Ghibaudo, T. Ouisse, E. Vogel, W. Hill, V. Misra, P. McLarty, and J. J. Wortman, Low-frequency noise characterization of n- and p-MOSFET's with ultrathin oxynitride gate films, *IEEE Electron Device Lett.* **17**, 395-397 (1996).
58. M. Da Rold, E. Simoen, S. Mertens, M. Schaekers, G. Badenes, and S. Decoutere, Impact of gate oxide nitridation process on 1/f noise in 0.18 μm CMOS, *Microelectron. Reliab.* **41**, 1933-1938 (2001).
59. M. Marin, J. C. Vildeuil, B. Tavel, B. Duriez, F. Arnaud, P. Stolk, and M. Woo, Can 1/f noise in MOSFETs be reduced by gate oxide and channel optimization?, in *Proc. Int. Conf. Noise and Fluctuations (ICNF)*, 2005, pp. 195-198.

60. T. Contaret, K. Romanjek, T. Boutchacha, G. Ghibaudo, and F. Bœuf, Low frequency noise characterization and modelling in ultrathin oxide MOSFETs, *Solid-State Electron.* **50**, 63-68 (2006).
61. J. Lee and G. Bosman, Comprehensive noise performance of ultrathin oxide MOSFETs at low frequencies, *Solid-State Electron.* **48**, 61-71 (2004).
62. J. Lee and G. Bosman, Defect spectroscopy using $1/f^\gamma$ noise of gate leakage current in ultrathin oxide MOSFETs, *Solid-State Electron.* **47**, 1973-1981 (2003).
63. J. Lee, G. Bosman, K. R. Green, and D. Ladwig, Model and analysis of gate leakage current in ultrathin nitrided oxide MOSFETs, *IEEE Trans. Electron Devices* **49**, 1232-1241 (2002).
64. J. Lee, G. Bosman, K. R. Green, and D. Ladwig, Noise model of gate-leakage current in ultrathin oxide MOSFETs, *IEEE Trans. Electron Devices* **50**, 2499-2506 (2003).
65. R. People, Physics and applications of Ge_xSi_{1-x}/Si strained-layer heterostructures, *IEEE J. Quantum Electronics* **QE-22**, 1696-1710 (1986).
66. D. J. Robbins, L. T. Canham, S. J. Barnett, A. D. Pitt, and P. Calcott, Near-band-gap photoluminescence from pseudomorphic $Si_{1-x}Ge_x$ single layers on silicon, *J. Appl. Phys.* **71**, 1407-1414 (1992).
67. T. Manku, J. M. McGregor, A. Nathan, D. J. Roulston, J.-P. Noel, and D. C. Houghton, Drift hole mobility in strained and unstrained doped $Si_{1-x}Ge_x$ alloys, *IEEE Trans. Electron Devices* **40**, 1990-1996 (1993).
68. T. Manku and A. Nathan, Effective mass for strained p-type $Si_{1-x}Ge_x$, *J. Appl. Phys.* **69**, 8414-8416 (1991).
69. S. Verdonckt-Vandebroek, E. F. Crabbé, B. S. Meyerson, D. L. Harame, P. J. Restle, J. M. C. Stork, and J. B. Johnson, SiGe-Channel Heterojunction p-MOSFET's, *IEEE Trans. Electron Devices*, **41**, 90-101 (1994).
70. C.-G. Ahn, H.-S. Kang, Y.-K. Kwon, and B. Kang, Effects of segregated Ge on electrical properties of SiO_2/SiGe interface, *Jpn. J. Appl. Phys.*, **37**, 1316-1319 (1998).
71. T. Ngai, X. Chen, J. Chen, and S. K. Banerjee, Improving SiO_2/SiGe interface of SiGe p-metal-oxide-semiconductor field-effect transistors using water vapor annealing, *Appl. Phys. Lett.* **80**, 1773-1775 (2002).
72. M. von Haartman, A.-C. Lindgren, P.-E. Hellström, M. Östling, T. Ernst, L. Brévard and S. Deleonibus, Influence of gate width on 50 nm gate length $Si_{0.7}Ge_{0.3}$ channel PMOSFETs, in *Proc. 33rd ESSDERC*, 2003, pp. 529-532.
73. N. Collaert, P. Verheyen, K. De Meyer, R. Loo, and M. Caymax, High performance Si/SiGe pMOSFETs fabricated in a standard CMOS process technology, *Solid-State Electron.* **47**, 1173-1177 (2003).
74. F. Andrieu, T. Ernst, K. Romanjek, O. Weber, C. Renard, J.-M. Hartmann, A. Toffoli, A.-M. Papon, R. Truche, P. Holliger, L. Brévard, G. Ghibaudo, and S. Deleonibus, SiGe channel p-MOSFETs scaling-down, in *Proc. ESSDERC*, 2003, pp. 267-270.
75. K. Romanjek, F. Andrieu, T. Ernst, and G. Ghibaudo, Characterization of the effective mobility by split $C(V)$ technique in sub 0.1 μm Si and SiGe PMOSFETs, *Solid-State Electron.* **49**, 721-726 (2005).
76. A.-C. Lindgren, P.-E. Hellberg, M. von Haartman, D. Wu, C. Menon, S.-L. Zhang, and M. Östling, Enhanced intrinsic gain (g_m/g_d) of PMOSFETs with a $Si_{0.7}Ge_{0.3}$ channel, in *Proc. ESSDERC*, 2002, pp. 175-178.
77. J. Alieu, T. Skotnicki, E. Josse, J.-L. Regolini, and G. Bremond, Multiple SiGe well: a new channel architecture for improving both NMOS and PMOS performances, in *Proc. Symp. VLSI Techology*, 2000, pp. 130-131.

78. Y.-C. Yeo, V. Subramanian, J. Kedzierski, P. Xuan, T.-J. King, J. Bokor, and C. Hu, Design and fabrication of 50-nm thin-body p-MOSFETs with a SiGe heterostructure channel, *IEEE Trans. Electron Devices*, **49**, 279-286 (2002).
79. T. Tsuchiya, T. Matsuura, and J. Murota, Low-Frequency Noise in $Si_{1-x}Ge_x$ p-Channel Metal Oxide Semiconductor Field-Effect Transistors, *Jpn. J. Appl. Phys. Part 1* **40**, 5290-5293 (2001).
80. P. W. Li, W. M. Liao, C. C. Shih, T. S. Kuo, L. S. Lai, Y. T. Tseng, and M. J. Tsai, High performance Si/SiGe heterostructure MOSFETs for low power analog circuit applications, *Solid-State Electron.* **47**, 1095-1098 (2003).
81. G. Ghibaudo and J. Chroboczek, On the origin of the LF noise in Si/Ge MOSFETs, *Solid-State Electron.* **46**, 393-398 (2002).
82. M. J. Prest, A. R. Bacon, D. J. F. Fulgoni, T. J. Grasby, E. H. C. Parker, T. E. Whall, and A. M. Waite, Low-frequency noise mechanisms in Si and pseudomorphic SiGe p-channel field-effect transistors, *Appl. Phys. Lett.* **85**, 6019-6021 (2004).
83. M. Myronov, O. A. Mironov, S. Durov, T. E. Whall, E. H. C. Parker, T. Hackbarth, G. Höck, H.-J. Herzog, and U. König, Reduced $1/f$ noise in p-$Si_{0.3}Ge_{0.7}$ metamorphic metal-oxide-semiconductor field-effect transistor, *Appl. Phys. Lett.* **84**, 610-612 (2004).
84. P. Srinivasan, E. Simoen, B. De Jaeger, C. Claeys, and D. Misra, $1/f$ noise performance of MOSFETs with HfO_2 and metal gate on Ge-on-insulator substrates, *Mat. Sci. Sem. Proc.* in press (2006).
85. M. V. Fischetti and S. E. Laux, Band structure, deformation potentials, and carrier mobility in strained Si, Ge and SiGe alloys, *J. Appl. Phys.* **80**, 2234-2252 (1996).
86. S. E. Thompson, M. Armstrong, C. Auth, M. Alavi, M. Buehler, R. Chau, S. Cea, T. Ghani, G. Glass, T. Hoffman, C.-H. Jan, C. Kenyon, J. Klaus, K. Kuhn, Z. Ma, B. Mcintyre, K. Mistry, A. Murthy, B. Obradovic, R. Nagisetty, P. Nguyen, S. Sivakumar, R. Shaheed, L. Shifren, B. Tufts, S. Tyagi, M. Bohr, and Y. El-Mansy, A 90-nm logic technology featuring strained-silicon, *IEEE Trans. Electron Devices* **51**, 1790-1797 (2004).
87. J. L. Hoyt, H. M. Nayfeh, S. Eguchi, I. Aberg, G. Xia, T. Drake, E. A. Fitzgerald, and D. A. Antoniadis, Strained silicon MOSFET technology, in *IEDM Tech. Dig.*, 2002, pp. 23-26.
88. S. H. Olsen, K. S. K. Kwa, L. S. Driscoll, S. Chattopadhyay, and A. G. O'Neill, Design, fabrication and characterisation of strained Si/SiGe MOS transistors, *IEE Proc.-Circuits Devices Syst.* **151**, 431-437 (2004).
89. T. Tezuka, N. Sugiyama, and S. Takagi, Fabrication of strained Si on an ultrathin SiGe-on-insulator virtual substrate with a high-Ge fraction, *Appl. Phys. Lett.* **79**, 1798-1800 (2001).
90. M. V. Fischetti, Z. Ren, P. M. Solomon, M. Yang, and K. Rim, Six-band k·p calculation of the hole mobility in silicon inversion layers: dependence on surface orientation, strain, and silicon thickness, *J. Appl. Phys.* **94**, 1079-1095 (2003).
91. M. Yang, M. Ieong, L. Shi, K. Chan, V. Chan, A. Chou, E. Gusev, K. Jenkins, D. Boyd, Y. Ninomiya, D. Pendleton, Y. Surpris, D. Heenan, J. Ott, K. Guarini, C. D'Emic, M. Cobb, P. Mooney, B. To, N. Rovedo, J. Benedict, R. Mo, and H. Ng, High performance CMOS fabricated on hybrid substrate with different crystal orientations, in *IEDM Tech. Dig.*, 2003, pp. 453-456.
92. T. Komoda, A. Oishi, T. Sanuki, K. Kasai, H. Yoshimura, K. Ohno, M. Iwai, M. Saito, F. Matsuoka, N. Nagashima, and T. Noguchi, Mobility improvement for 45 nm node by combination of optimized stress control and channel orientation design, in *IEDM Tech. Dig.*, 2004, pp. 217-220.

93. K. Rim, J. Chu, H. Chen, K. A. Jenkins, T. Kanarsky, K .Lee, A. Mocuta, H. Zhu, R. Roy, J. Newbury, J .Ott, K. Petrarca, P. Mooney, D. Lacey, S. Koester, K. Chan, D. Boyd, M. Ieong, and H.-S. Wong, Characteristics and device design of sub-100 nm strained Si N- and PMOSFETs, in *Proc. Symp. VLSI Technology*, 2002, pp. 98-99.
94. F. Lime, F. Andrieu, J. Derix, G. Ghibaudo, F. Boeuf, and T. Skotnicki, Low temperature characterization of effective mobility in uniaxially and biaxially strained nMOSFETs, *Solid-State Electron.* **50**, 644-649 (2006).
95. J.-S. Goo, Q. Xiang, Y. Takamura, H. Wang; J. Pan, F. Arasnia, E. N. Paton, P. Besser, M. V. Sidorov, E. Adem, A. Lochtefeld, G. Braithwaite, M .T. Currie, R. Hammond, M. T. Bulsara, and M.-R. Lin, Scalability of strained-Si nMOSFETs down to 25 nm gate length, *IEEE Electron Device Letters* **24**, 351-353 (2003).
96. T. Ohguro, Y. Okayama, K. Matsuzawa, K. Matsunaga, N. Aoki, K. Kojima, H. S. Momose, and K. Ishimaru, The impact of oxynitride process, deuterium annealing and STI stress to 1/f noise of 0.11 μm CMOS, in *Proc. Symp. VLSI Technology*, 2003, pp. 37-38.
97. G. Giusi, E. Simoen, G. Eneman, P. Verheyen, F. Crupi, K. De Meyer, C. Claeys, and C. Ciofi, Low-frequency (1/*f*) noise behavior of locally stressed HfO_2/TiN gate-stack pMOSFETs, *IEEE Electron Device Lett.* **27**, 508-510 (2006).
98. M. H. Lee, P. S. Chen, W.-C. Hua, C.-Y. Yu, Y. T. Tseng, S. Maikap, Y. M. Hsu, C. W. Liu, S. C. Lu, W.-Y. Hsieh, and M.-J. Tsai, Comprehensive low-frequency and RF noise characteristics in strained-Si NMOSFETs, in *IEDM Tech. Dig.*, 2003, pp. 69-72.
99. W.-C. Hua, M. H. Lee, L. P. S. Chen, S. Maikap, C. W. Liu, and K. M. Chen, Ge outdiffusion effect on flicker noise in strained-Si nMOSFETs, *IEEE Electron Device Lett.* **25**, 693-695 (2004).
100. W.-C. Hua, M. H. Lee, P. S. Chen, M.-J. Tsai, and C. W. Liu, Threading dislocation induced low frequency noise in strained-Si nMOSFETs, *IEEE Electron Device Lett.* **26**, 667-669 (2005).
101. E. Simoen, G. Eneman, P. Verheyen, R. Loo, K. De Meyer, and C. Claeys, Processing aspects in the low-frequency noise of nMOSFETs on strained-silicon substrates, *IEEE Trans. Electron Devices* **53**, 1039-1047 (2006).
102. J.-P. Colinge, Fully-depleted SOI CMOS for analog applications, *IEEE Trans. Electron Devices* **45**, 1010-1016 (1998).
103. S. Cristoloveanu, Silicon on insulator technologies and devices: from present to future, *Solid-State Electron.* **45**, 1403-1411 (2001).
104. T. Ernst, S. Cristoloveanu, G. Ghibaudo, T. Ouisse, S. Horiguchi, Y. Ono, Y. Takahashi, and K. Murase, Ultimately thin double-gate SOI MOSFETs, *IEEE Trans. Electron Devices* **50**, 830-838 (2003).
105. S. Narasimha, A. Ajmera, H. Park, D. Schepis, N. Zamdmer, K. A. Jenkins, J.-O. Plouchart, W.-H. Lee, J. Mezzapelle, J. Bruley, B. Doris, J. W. Sleight, S. K. Fung, S. H. Ku, A. C. Mocuta, I. Yang, P. V. Gilbert, K. P. Muller, P. Agnello, and J. Welser, High performance sub-40nm CMOS devices on SOI for the 70nm technology node, in *IEDM Tech. Dig.*, 2001, pp. 625-628.
106. S. Monfray, T. Skotnicki, C. Fenouillet-Beranger, N. Carriere, D. Chanemougame, Y. Morand, S. Descombes, A. Talbot, D. Dutartre, C. Jenny, P. Mazoyer, R. Palla, F. Leverd, Y. Le Friec, R. Pantel, S. Borel, D. Louis, and N. Buffet, Emerging silicon-on-nothing (SON) devices technology, *Solid-State Electron.* **48**, 887-895 (2004).
107. B .Doris, M. Ieong, T. Kanarsky, Y. Zhang, R. A. Roy, O. Dokumaci, Z. Ren, F.-F. Jamin, L. Shi, W. Natzle, H.-J. Huang, J. Mezzapelle, A. Mocuta, S. Womack, M. Gribelyuk, E. C. Jones, R. J. Miller, H.-S. P. Wong, and W. Haensch, Extreme scaling with ultra-thin Si channel MOSFETs, in *IEDM Tech. Dig.*, 2002, pp. 267-270.

4. 1/f noise performance of advanced CMOS devices

108. M. Bruel, B. Aspar, B. Charlet, C. Maleville, T. Poumeyrol, A. Soubie, A. J. Auberton-Herve, J. M. Lamure, T. Barge, F. Metral, and S. Trucchi, 'Smart cut': a promising new SOI material technology, in *Proc. IEEE Int. SOI Conf.*, 1995, pp. 178-179.
109. C. Claeys, E. Simoen, A. Efremov, V. G. Litovchenko, A. Evtukh, A. Kizjak, and Ju. Rassamakin, γ-irridation hardness of short-channel nMOSFETs fabricated in a 0.5 μm SOI technology, *Nucl. Instr. and Meth. B* **186**, 429-434 (2002).
110. A. Mercha, E. Simoen, H. van Meer, and C. Claeys, Low-frequency noise overshoot in ultrathin gate oxide silicon-on-insulator metal-oxide-semiconductor field-effect-transistors, *Appl. Phys. Lett.* **82**, 1790-1792 (2003).
111. F.-L. Yang, H.-Y. Chen, F.-C. Chen, C.-C. Huang, C.-Y. Chang, H.-K. Chiu, C.-C. Lee, C.-C. Chen, H.-T. Huang, C.-J. Chen, H.-J. Yeo, M.-S. Liang, and C. Hu, 25 nm CMOS Omega FETs, in *IEDM Tech. Dig.*, 2002, pp. 255-258.
112. J.-P. Colinge, Conduction mechanisms in thin-film accumulation-mode SOI p-channel MOSFET's, *IEEE Trans. Electron Devices* **37**, 718-723 (1990).
113. E. Simoen, and C. Claeys, The low-frequency noise behaviour of silicon-on-insulator technologies, *Solid-State Electron.* **39**, 949-960 (1996).
114. M. Matloubian, F. Scholz, and L. Lum, Low frequency noise in fully depleted SOI PMOSFET's, *IEEE Trans. Electron Devices* **41**, 1977-1980 (1994).
115. N. Lukyanchikova, M. Petrichuk, N. Garbar, E. Simoen, and C. Claeys, Back and front interface related generation-recombination noise in buried-channel SOI pMOSFETs, *IEEE Trans. Electron Devices* **43**, 417-423 (1996).
116. F. Balestra, S. Cristoloveanu, M. Benachir, J. Brini, and T. Elewa, Double-gate silicon-on-insulator transistor with volume inversion: a new device with greatly enhanced performance, *IEEE Electron. Device Lett.* **EDL-8**, 410-412 (1987).
117. Y.-C. Tseng, W. M. Huang, M. Mendicino, D. J. Monk, P. J. Welch, and J. C. S. Woo, Comprehensive study on low-frequency noise characteristics in surface channel SOI CMOSFETs and device design optimization for RF ICs, *IEEE Trans. Electron Devices*, **48**, 1428-1437 (2001).
118. Web-based simulation tool, http://www.nanohub.org
119. J. Hållstedt, M. von Haartman, P.-E. Hellström, M. Östling and H. H. Radamson, Hole mobility in ultrathin body SOI pMOSFETs with SiGe or SiGeC channels, *IEEE Electron Device Lett* **27**, 466-468, 2006.
120. M. Fritze, C. L. Chen, S. Calawa, D. Yost, B. Wheeler, P. Wyatt, C. L. Keast, J. Snyder, and J. Larson, High-speed schottky-barrier pMOSFET with f_T = 280 GHz, *IEEE Electron Device Lett.* **25**, 220-222 (2004).
121. J. Seger, P.-E. Hellström, J. Lu, B. G. Malm, M. von Haartman, M. Östling, and S.-L. Zhang, Lateral encroachment of Ni-silicides in the source/drain regions on ultrathin silicon-on-insulator, *Appl. Phys. Lett.* **86**, 253507 (2005).
122. K. M. Cao, W.-C. Lee, W. Liu, X. Jin, P. Su, S. K. H. Fung, J. X. An, B. Yu, and C. Hu, BSIM4 gate leakage model including source-drain partition, in *IEDM Tech. Dig.*, 2000, pp. 815-818.
123. G. D. Wilk, R. M. Wallace, and J. M. Anthony, High-κ gate dielectrics: Current status and materials properties considerations, *J. Appl. Phys.* **89**, 5243-5275 (2001).
124. E. P. Gusev, D. A. Buchanan, E. Cartier, A. Kumar, D. DiMaria, S. Guha, A. Callegari, S. Zafar, P. C. Jamison, D. A. Neumayer, M. Copel, M. A. Gribelyuk, H. Okron-Schmidt, C. D'Emic, P. Kozlowski, K. Chan, N. Bojarczuk, L.-Å. Ragnarsson, P. Ronsheim, K. Rim, R. J. Fleming, A. Mocuta, and A. Ajmera, Ultrathin high-K gate stacks for advanced CMOS devices, in *IEDM Tech. Dig.*, 2001, pp. 451-454.
125. A. L. P. Rotondaro, M. R. Visokay, J. J. Chambers, A. Shanware, R. Khamankar, H. Bu, R. T. Laaksonen, L. Tsung, M. Douglas, R. Kuan, M. J. Bevan, T. Grider, J. McPherson,

and L. Colombo, Advanced CMOS transistors with a novel HfSiON gate dielectric, in *Proc. Symp. VLSI Technology*, 2002, pp. 148-149.
126. W. J. Zhu and T. P. Ma, Temperature Dependence of channel mobility in HfO_2-gated NMOSFETs, *IEEE Electron Device Lett* **25**, 89-91 (2004).
127. S. Saito, D. Hisamoto, S. Kimura, and M. Hiratani, Unified mobility model for high-κ gate stacks, in *IEDM Tech. Dig.*, 2003, pp.797-800.
128. Z. Ren, M. V. Fischetti, E. P. Gusev, E. A. Cartier, and M. Chudzik, Inversion channel mobility in high-κ high performance MOSFETs, in *IEDM Tech Dig.*, 2003, pp. 793-796.
129. L.-Å. Ragnarsson, L. Pantisano, V. Kaushik, S.-I. Saito, Y. Shimamoto, S. De Gendt, and M. Heyns, The impact of sub monolayers of HfO_2 on the device performance of high-k based transistors, in *IEDM Tech. Dig.*, 2003, pp. 87-90.
130. G.-W. Lee, J.-H. Lee, H.-W. Lee, M.-K. Park, D.-G. Kang, and H.-K. Youn, Trap evolutions of metal/oxide/silicon field-effect transistors with high-*k* gate dielectric using charge pumping method, *Appl. Phys. Lett.* **81**, 2050-2052 (2002).
131. J. Robertson, Interfaces and defects of high-*K* oxides on silicon, *Solid-State Electron.* **49**, 283-293 (2005).
132. M. Fadlallah, A. Szewczyk, C. Giannakopoulos, B. Cretu, F. Monsieur, T. Devoivre, J. Jomaah, and G. Ghibaudo, Low-frequency noise and reliability properties of 0.12 μm CMOS devices with Ta_2O_5 as gate dielectrics, *Microelectron. Reliab.* **41**, 1361-1366, (2001).
133. T. Ishikawa, S. Tsujikawa, S. Saito, D. Hisamoto, and S. Kimura, Direct evaluation of an interfacial layer in high-κ gate dielectrics by $1/f$ noise measurements, in *Proc. Int. Conf. Solid State Devices and Materials (SSDM)*, 2003, pp. 14-15.
134. E. Simoen, A. Mercha, L. Pantisano, C. Claeys, and E. Young, Low-frequency noise study of n-MOSFETs with HfO_2 gate dielectric, in *Proc. Vol. 22 Electrochem Soc. Meeting*, 2003, pp. 319-331.
135. H. Sauddin, Y. Yoshihara, S. Ohmi, K. Tsutsui, and H. Iwai, Low-frequency noise characteristics of MISFETs with La_2O_3 gate dielectrics, in *Proc. Vol. 22 Electrochem Soc. Meeting*, 2003, pp. 415-423.
136. H. D. Xiong, D. M. Fleetwood, J. A. Felix, E. P. Gusev, and C. D'Emic, Low-frequency noise and radiation response of metal-oxide-semiconductor transistors with $Al_2O_3/SiO_xN_y/Si(100)$ gate stacks, *Appl. Phys. Lett.* **83**, 5232-5234 (2003).
137. E. Simoen, A. Mercha, L. Pantisano, C. Claeys, and E. Young, Low-frequency noise behavior of SiO_2-HfO_2 dual-layer gate dielectric nMOSFETs with different interfacial oxide thickness, *IEEE Trans. Electron Devices* **51**, 780-784 (2004).
138. A. Kerber, E. Cartier, L. Pantisano, R. Degraeve, T. Kauerauf, Y. Kim, A. Hou, G. Groeseneken, H. E. Maes, and U. Schwalke, Origin of the threshold voltage instability in SiO_2/HfO_2 dual layer gate dielectrics, *IEEE Electron Device Lett.* **24**, 87-89 (2003).
139. S. Zafar, A. Callegari, E. Gusev, and M. V. Fischetti, Charge trapping related threshold voltage instabilities in high permittivity gate dielectric stacks, *J. Appl. Phys.* **93**, 9298-9303 (2003).
140. J. S. Cable, R. A. Mann, and J. C. S. Woo, Impurity barrier properties of reoxidized nitrided oxide films for use with P+-doped polysilicon gates, *IEEE Electron Device Lett.*, **12**, 128-130 (1991).
141. C. Claeys, E. Simoen, A. Mercha, L. Pantisano, and E. Young, Low-frequency noise performance of HfO_2-based gate stacks, *J. Electrochem. Soc.* **152**, F115-F123 (2005).
142. D. Wu, A.-C. Lindgren, S. Persson, G. Sjöblom, M. von Haartman, J. Seger, P.-E. Hellström, J. Olsson, H.-O. Blom, S.-L. Zhang, M. Östling, E. Vainonen-Ahlgren, W.-M. Li, E. Tois, and M. Tuominen, A novel strained $Si_{0.7}Ge_{0.3}$ surface-channel pMOSFET with

an ALD TiN/Al$_2$O$_3$/HfAlO$_x$/Al$_2$O$_3$ gate stack, *IEEE Electron Device Lett.* **24**, 171-173 (2003).
143. M. V. Fischetti, D. A. Neumayer, and E. A. Cartier, Effective electron mobility in Si inversion layers in metal-oxide-semiconductor systems with a high-κ insulator: The role of remote phonon scattering, *J. Appl. Phys.* **90**, 4587-4608 (2001).
144. E. Simoen, A. Mercha, L. Pantisano, C. Claeys, and E. Young, Tunneling 1/$f^γ$ noise in 5 nm HfO$_2$/2.1 nm SiO$_2$ gate stack n-MOSFETs, *Solid-State Electron.* **49**, 702-707 (2005).
145. C. Leroux, J. Mitard, G. Ghibaudo, X. Garros, G. Reimbold, B. Guillaumot, and F. Martin, Characterization and modelling of hysteresis phenomena in high k dielectrics, in *IEDM Tech. Dig.*, 2004, pp. 737-740.
146. H .Yu, Y.-T. Hou, M.-F. Li, and D.-L. Kwong, Investigation of hole-tunneling current through ultrathin oxynitride/oxide stack gate dielectrics in p-MOSFETs, *IEEE Trans. Electron Devices* **49**, 1158-1164 (2002).
147. H.-H. Tseng, C. C. Capasso, J. K. Schaeffer, E. A. Hebert, P. J. Tobin, D. C. Gilmer, D. Triyoso, M. E. Ramón, S. Kalpat, E. Luckowski, W. J. Taylor, Y. Jeon, O. Adetutu, R. I. Hedge, R. Noble, M. Jahanbani, C. El-Chemali, and B.White, Improved short channel device characteristics with stress relieved pre-oxide (SRPO) and a novel tantalum carbon alloy metal gate/HfO$_2$ stack, in *IEDM Tech. Dig.*, 2004, pp. 821-824.
148. B. Min, S. P. Devireddy, Z. Çelik-Butler, A. Shanware, L. Colombo, K. Green, J. J. Chambers, M. R. Visokay, and A. L. P. Rotondaro, Impact of interfacial layer on low-frequency noise of HfSiON dielectric MOSFETs, *IEEE Trans. Electron Devices* **53**, 1459-1466 (2006).
149. M. Sandén, B. Gunnar Malm, J. V. Grahn, and M. Östling, Decreased low-frequency noise by hydrogen passivation of polysilicon emitter bipolar transistors, *Microelectron. Reliab.* **40**, 1863-1867 (2000).
150. P. Srinivasan, E. Simoen, Z. M. Rittersma, W. Deweerd, L. Pantisano, C. Claeys, and D. Misra, Effect of nitridation on low-frequency (1/f) noise in n- and p-MOSFETs with HfO$_2$ gate dielectrics, *J. Electrochem. Soc.* **153**, G819-G825 (2006).
151. M. von Haartman, J. Westlinder, D. Wu, B. G. Malm, P.-E. Hellström, J. Olsson and M. Östling, Low-frequency noise and Coulomb scattering in Si$_{0.8}$Ge$_{0.2}$ surface channel pMOSFETs with ALD Al$_2$O$_3$ gate dielectrics, *Solid-State Electronics* **49**, 907-914 (2005).
152. P. Xuan and J. Bokor, Investigation of NiSi and TiSi as CMOS gate materials, *IEEE Electron Device Lett.* **24**, 634-636 (2003).
153. G. Sjöblom, J. Westlinder, and J. Olsson, Investigation of the thermal stability of reactively sputter deposited TiN MOS gate electrodes, *IEEE Trans. Electron Devices* **52**, 2349-2352 (2005).
154. R. Chau, S. Datta, M. Doczy, B. Doyle, J. Kavalieros, and M. Metz, High-κ/metal-gate stack and its MOSFETs characteristics, *IEEE Electron Device Lett.* **25**, 408-410 (2004).
155. P. Srinivasan, E. Simoen, L. Pantisano, C. Claeys, and D. Misra, Impact of gate material on low-frequency noise of nMOSFETs with 1.5 nm SiON gate dielectric: testing the limits of the number fluctuation theory, in *Proc. Int. Conf. Noise and Fluctuations (ICNF)*, 2005, pp. 231-234.
156. D. Esseni, M. Mastrapasqua, C. Fiegna, G. K. Keller, L. Selmi, and E. Sangiorgi, An experimental study of low-field electron mobility in double-gate ultra-thin SOI MOSFETs, in *IEDM Tech. Dig.*, 2001, pp. 445-448.
157. E. Simoen, U. Magnusson, and C. Claeys, A low-frequency noise study of gate-all-around SOI transistors, *IEEE Trans. Electron Devices* **40**, 2054-2059 (1993).
158. B. Yu, L. Chang, S. Ahmed, H. Wang, S. Bell, C.-Y. Yang, C. Tabery, C. Ho, Q. Xiang, T.-J. King, J. Bokor, C. Hu, M.-R. Lin, and D. Kyser, FinFET scaling to 10nm gate length, in *IEDM Tech. Dig.*, 2002, pp. 251-254.

159. D. Esseni, A. Abramo, L. Selmi, and E. Sangiorgi, Physically based modeling of low field electron mobility in ultrathin single- and double-gate SOI nMOSFETs, *IEEE Trans. Electron Devices* **50**, 2445-2455 (2003).
160. P. Gaubert, A. Teramoto, T. Hamada, M. Yamamoto, K. Kotani, and T. Ohmi, $1/f$ noise suppression of pMOSFETs fabricated on Si(110) and Si(100) using an alkali-free cleaning process, *IEEE Trans. Electron Devices* **53**, 851-856, 2006.
161. H. S. Momose, T. Ohguro, K. Kojima, S.-I. Nakamura, and Y. Toyoshima, 1.5-nm gate oxide CMOS on (110) surface-oriented Si substrate, *IEEE Trans. Electron Devices* **50**, 1001-1008 (2003).
162. J.-S. Lee, Y.-K. Choi, D. Ha, T.-J. King, and J. Bokor, Low-frequency noise characteristics in p-channel FinFETs, *IEEE Electron Device Lett.* **23**, 722-724 (2002).
163. K. Akarvardar, B. M. Dufrene, S. Cristoloveanu, P. Gentil, B. J. Blalock, and M. M. Mojarradi, Low-frequency noise in SOI four-gate transistors, *IEEE Trans. Electron Devices* **53**, 829-835 (2006).

PROBLEMS

1. The drain current noise is measured in a Si nMOSFET biased at $V_{GS} - V_T = 0.1$ V and $V_{DS} = 25$ mV. The drain current noise in the channel is found to be of the $1/f$ type and equals

$$S_{I_D} = 1.4 \times 10^{-17} \text{ A}^2/\text{Hz at } f = 10 \text{ Hz and } I_D = 114 \text{ } \mu\text{A}.$$

(a) Estimate the oxide trap density and Hooge parameter for the device.
(b) Does the device fulfil the ITRS requirements on the $1/f$ noise?

Device parameters:
$W = 10$ µm
$L = 75$ nm
$t_{EOT} = 2.2$ nm (SiO$_2$).

2. A new device is being developed that will triple the cut-off frequency f_T. The device can be designed in four different CMOS technologies. Evaluate these alternatives from a gate voltage noise point of view and compare with the original device. Number fluctuations dominate unless otherwise mentioned.

i) The device gate length L and t_{EOT} are scaled by a factor 1/2. A high-k gate dielectric is used in order to be able to scale down the t_{EOT} that much, however at the expense at a six-fold increase in the oxide trap density. The mobility decreases to 3/4 of its original value in this device.

ii) Strain engineering is used to enhance the mobility by 69%. The gate length and t_{EOT} are scaled by a factor 3/4. In order to scale down t_{EOT}, the nitrogen content in the gate oxide was increased. The 1/f noise was found to be dominated by mobility fluctuations with a Hooge parameter value that was two times higher than that in the reference device.

iii) A fully depleted SOI-technology is used. L is scaled by a factor 2/3. The body coefficient decreases 25% and the mobility remains unchanged. The front oxide trap density is the same as in the reference device, but the back oxide trap density is three times higher than in the front.

iv) A new advanced channel material that improves the mobility by a factor of three was developed for a device with a 2 times larger t_{EOT} than in the reference device. Both the oxide trap density and the gate length are unchanged.

Chapter 5

INTRODUCTION TO NOISE IN RF/ANALOG CIRCUITS

1. INTRODUCTION

Up to now, we have discussed the low-frequency (LF) noise in MOS transistors, the different noise mechanisms, models and various technology considerations. One intriguing questions remains, how does the LF noise affect an electronic circuit? LF noise is presently not an important consideration for a digital circuit, as can be understood intuitively, since discrete voltage levels (represented by "zeroes" and "ones") are processed. The digital circuits are primarily optimized by the trade-off between speed and power consumption. For an RF/analog circuit on the other hand, several additional circuit properties (including noise) must be considered. Thus, the design of RF and analog circuits involve considerations and trade-offs such as gain, power dissipation, linearity, noise, speed, voltage swings, input/output impedance and supply voltage.[1] LF noise is one of the key difficulties in RF and analog circuit design, which is expected to increase in importance as the devices are downsized and the technology becomes more and more advanced. At the 45 nm node and beyond the LF noise could even affect the function of digital circuits as predicted by Deen and Marinov.[2] The complexity of RF design entails specialized characterization and accurate modeling of the RF devices,[1,3] which explains our efforts in the previous chapters to discuss these issues. In this chapter, we will explain why noise, and in particular, LF noise is a problem in analog/RF circuits. Some of the most important circuits, the voltage controlled oscillator (VCO), mixer and low-noise amplifier (LNA), are discussed from a noise point of view. These circuits are found in transmitters-receivers (transceivers) systems used in communications.

Someone familiar with electronic circuits may find the presentation here too simplified, but the primary goal is to provide an introduction to noise in analog/RF circuits in order to bring together the physics and device world with the circuit world. We mainly discuss the effects of LF noise and thermal noise on the above mentioned electrical circuits. So called "noise coupling" effects, for example the coupling of power supply and transistor switching disturbances through the substrate or metal layers, are not treated here.

2. IMPACT OF $1/f$ NOISE IN RF CIRCUITS

The signal-to-noise ratio (SNR), equal to the signal power divided by the noise power, for an electronic circuit must be sufficiently high in order for the circuit to function properly. When the SNR is too low the desired signal becomes difficult to distinguish from the background noise causing degraded quality of the information delivered by the circuit, or in the worst case the circuit completely fails to deliver any intelligent information. In some cases, this can be circumvented by increasing the signal power, of course at the expense of increasing power dissipation. The total noise power should be integrated over the bandwidth of the circuit, as explained in chapter 1. Therefore, another way to increase the SNR is to limit the bandwidth. In this way, the effect of the $1/f$ noise can be avoided in many circuits. However, the $1/f$ noise might still contribute to the total noise power since it can be *upconverted* to higher frequencies. In a circuit with a periodically varying operating point, the noise can be modulated by the input signal causing *cyclostationary noise*. This occurs due to the noise sources being bias dependent or due to a modulation of the transfer function for the noise from the source to the output (nonlinearity).[4] Both the amplitude and the phase of the periodic signal can be modulated by noise, therefore one speaks about amplitude modulation (AM) and phase modulation (PM) noise. A simple example of cyclostationary noise is given in Fig. 5-1, showing a voltage-controlled switch being operated by a periodic signal. The periodic switching causes periodically varying fluctuations at the output.

Due to the upconversion of the $1/f$ noise, for example to phase noise in a VCO which will be described in the next section, the $1/f$ noise stemming from the devices shows up at higher frequencies and may cause a severe degradation of the SNR in the bandwidth of interest. Fig. 5-2 describes a receiver system where the bandwidth is divided in frequency slots for the different communication channels. The phase noise generated around the

5. Introduction to noise in RF/analog circuits

carrier frequency in an adjacent channel interferes with the signal in the channel of interest. The signal-to-noise ratio becomes degraded due to the phase noise. The influence of the phase noise can be reduced by placing the channels further apart in frequency. However, this obviously leads to fewer channels within the bandwidth resulting in lower information capacity of the communication system.

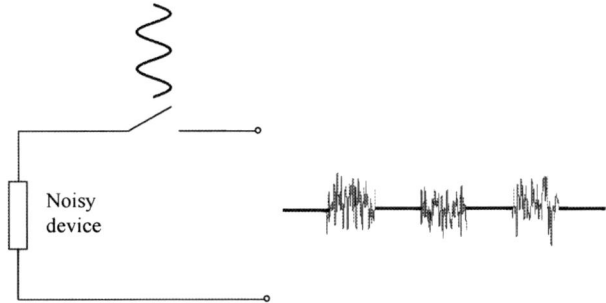

Figure 5-1. Example of cyclostationary noise. The voltage-controlled switch modulates the noise from the device.

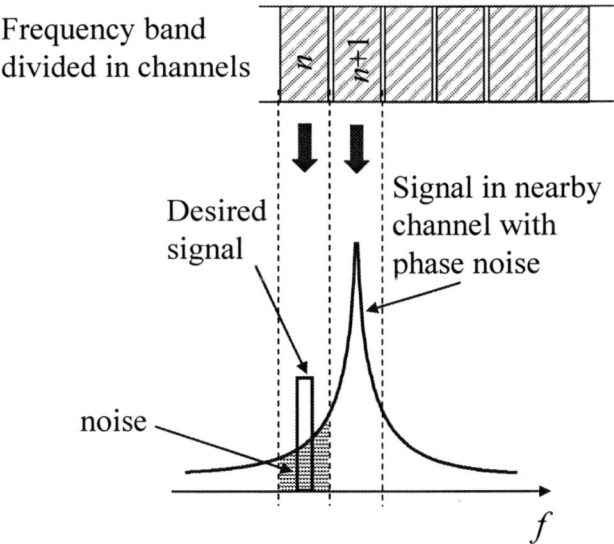

Figure 5-2. Description how phase noise could affect a receiver system. The phase noise around an interferer in an adjacent channel mixes with the signal in the desired channel and degrades the SNR.

3. VOLTAGE CONTROLLED OSCILLATOR (VCO)

3.1 The VCO and phase noise

Voltage controlled oscillators are used to produce a periodic signal at a certain frequency that can be varied by an applied voltage. A typical *LC* oscillator consists of a passive *LC*-tank whose resonance frequency sets the frequency of oscillation. The energy loss in the *LC*-tank is precisely compensated by the energy supplied by an active device, typically a transistor. The VCO is a key building block in wireless transceivers, where it is used together with the mixer to perform frequency translation. The received RF signal (at 900 or 1800 MHz in the GSM, for example) is multiplied with the oscillator signal in the mixer to downconvert the RF signal to an intermediate frequency (heterodyne architecture) or directly to the baseband (homodyne). The VCO and the mixer are used in the opposite way to generate an RF signal from the baseband signal in the transmit path. An ideal oscillator generates a perfect sinusoidal signal, which corresponds to a pulse in the frequency spectrum, see Fig. 5-3(a). For an actual oscillator, however, the spectrum exhibits "skirts" around the centre frequency called phase noise,[5] as shown in Fig. 5-3(b). Phase noise is a difficult problem in wireless transceivers; RF oscillators must therefore meet stringent phase noise requirements for these kinds of circuits. Both the desired signal and the signals in the adjacent channels are downconverted by the oscillator which exhibits finite phase noise. The phase noise from the downconverted interfering signal is mixed with the downconverted desired signal at the output, the two spectra overlap corrupting the signal-to-noise ratio, as explained in Fig. 5-2.

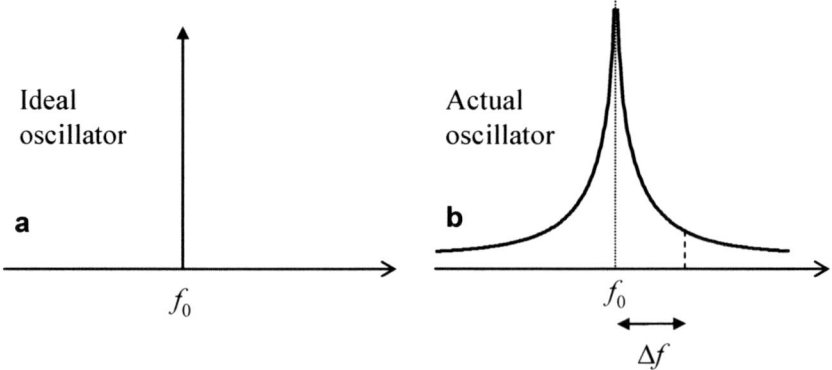

Figure 5-3. (a) Ideal oscillator signal. (b) Actual oscillator signal with phase noise.

3.2 Upconverted 1/f noise

Device 1/f noise is a particular problem for VCOs since it is upconverted to phase noise at small frequency offsets from the carrier frequency and therefore sets the ultimate separation limit of two channels.[5-9] Fig. 5-4 schematically illustrates the phase noise spectrum and the different physical origins of the phase noise. One drawback with oscillators implemented in CMOS compared to bipolar technology is the inferior 1/f noise performance in the former technology, which has been thought to exclude CMOS to be used in high-performance oscillators.[6] This is a motivation to study how 1/f noise is upconverted to phase noise in oscillators and to understand the mechanisms behind the 1/f noise in CMOS transistors in order to be able to reduce the phase noise originating from device 1/f noise by proper design.

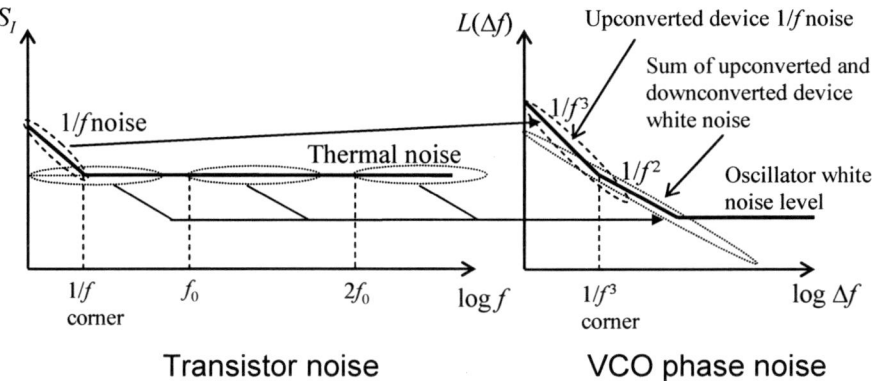

Figure 5-4. Schematic illustration of the phase noise spectrum (in log-log plot) and how the device noise is transferred to phase noise.

The 1/f noise of each transistor in the oscillator can contribute to the phase noise, but the transistors used for the frequency control are particularly important.[5] Fig. 5-5 shows an example of a complementary cross-coupled LC oscillator for low phase noise. The tail transistor (bottom transistor in Fig. 5-5) has been shown to be the main source of the upconverted 1/f noise.[10,11] The frequency of oscillation is a function of the current flowing through this device. Thus, low-frequency noise in the current is directly translated to low-frequency noise in the frequency of oscillation, to phase noise. The tail transistor should therefore be designed for low noise, for example by increasing the area.

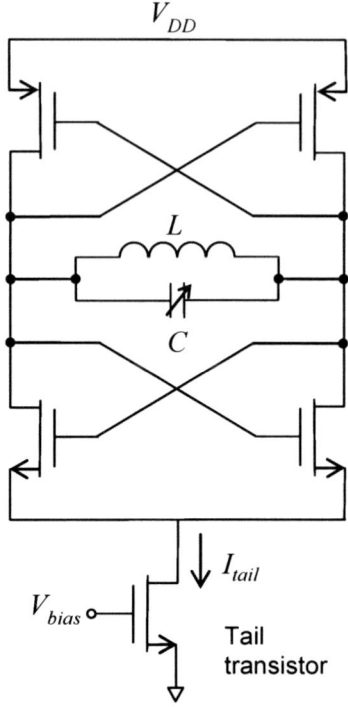

Figure 5-5. Complementary oscillator topology for low phase noise.

The phase noise generation also depends on the circuit symmetry. A general theory of phase noise in electrical oscillators based on linearization of a time variant system was given by Hajimiri and Lee.[6] Noise located near integer multiples of the oscillation frequency contributes to the total phase noise, according to their approach. The upconversion of the $1/f$ noise is sensitive to symmetry properties of the oscillator waveform. The impulse sensitivity function (ISF) describes how much phase shift that results from an impulse at different positions in the oscillation cycle. The ISF is periodic and can therefore be expanded in a Fourier series

$$ISF(\omega_0 t) = c_0/2 + \sum_{n=1}^{\infty} c_n \cos(n\omega_0 t + \theta_n). \tag{5-1}$$

Here, ω_0 is the frequency of oscillation, t is the time, c_n are Fourier coefficients. θ_n is the phase of the nth harmonic, which has turned out to be unimportant for the calculations and will therefore be ignored. The phase noise resulting from the transistor drain current noise can then be written

5. Introduction to noise in RF/analog circuits

$$L(\Delta f) = 10 \cdot \log\left(\frac{S_{I_{D,1/f}}(\Delta f) \cdot c_0^2 + \sum_{n=0}^{\infty} c_n^2 S_{I_{D,th}}(n \cdot f_0 + \Delta f)}{32\pi^2 q_{max}^2 \Delta f^2}\right)$$

$S_{I_{D,1/f}}$ = PSD of the $1/f$ noise in the drain current (5-2)

$S_{I_{D,th}}$ = PSD of the thermal noise in the drain current

where $q_{max} = C \times V_{max}$ is the maximum charge displacement across the tank capacitor C determined by the maximum voltage swing across the tank V_{max}. As seen in Eq. (5-2), the oscillator phase noise can be reduced by maximizing the voltage swing V_{max} and minimizing the coefficients c_n. Maximizing the quality factor (Q value) of the LC-tank will reduce the generated thermal noise and thereby lower the $1/f^2$-shaped phase noise. The DC value of the ISF, c_0, can actually be minimized by symmetry considerations. The VCO circuit presented in Fig. 5-5 provides good symmetry properties and therefore good phase noise performance. The relative widths of the nMOS and pMOS transistors must be selected in an appropriate way in order to minimize c_0 (the optimum width ratio may differ from the ratio that give equal transconductance). In this way, it is possible to reduce the $1/f^3$ phase noise corner to a few tens of kHz for a device with a $1/f$ noise corner of a few hundred kHz. A $1/f^3$ corner frequency of 15 kHz was for example reported in the works by De Muer et al.[10] and Kao et al.[12]

3.3 Phase noise performance

The phase noise is usually measured in dB relative to the carrier power with the unit dBc/Hz. The phase noise is often quoted at an offset frequency from the oscillation frequency of the oscillator, for example 600 kHz. Typical VCO performance characteristics in terms of oscillation frequency, power and phase noise are summarized in Table 5-1 for some published works. A commonly used VCO performance figure-of-merit defined as[12]

$$FOM = L(\Delta f) - 20\log(f_0/\Delta f) + 10\log(P_D/1\,\text{mW}) \quad [\text{dBc/Hz}] \quad (5-3)$$

weighs these performance criteria together and can be calculated from the data in Table 5-1.

Another type of oscillator, called the ring oscillator, can be realized by connecting an odd number of inverters in series forming a ring. The ring oscillator is compact and easy to integrate. It is the most fabricated of all oscillators and is used in various communications electronic circuits as well as a test vehicle for the semiconductor process.[13] However, its RF performance is poor due to high phase noise,[7] as observed in Table 5-1. For a more detailed analysis of ring oscillators, please refer to the recent paper by Abidi.[13]

Table 5-1. Performance overview of state-of-the-art VCOs.

Technology	Frequency f_0 (GHz)	Phase noise L (dBc/Hz)	Power dissipation P_D (mW)
LC VCO 0.35μm CMOS[12]	2.06 GHz	-116 @ 0.6 MHz	22.6
LC VCO 0.18μm CMOS[14]	12	-102 @ 0.6 MHz	7.7
LC VCO 0.65μm CMOS[10]	2	-125 @ 0.6 MHz	34
LC VCO 0.25μm CMOS[15]	1.8	-121 @ 0.6 MHz	6
3-stage Ring VCO 0.18μm CMOS[16]	9.5	-92 @ 1 MHz	61
2-stage Ring VCO 0.18μm CMOS[17]	3.6	-90 @ 1 MHz	17

4. MIXER

A mixer contains two inputs, one for the information carrying signal and one for the local oscillator (LO) signal, and performs frequency translation by multiplying these signals. The signals generated at the mixer output have frequencies corresponding to the sum and difference of the signals at the input, respectively. When the mixer is used in a transmitter, the low-frequency baseband signal is upconverted to RF frequencies by the mixer. A mixer employed in a receiver performs the opposite translation by downconverting the received RF signal to intermediate frequencies (IF) for further processing. A receiver architecture that converts the RF signal directly the baseband ("zero" IF) is called a homodyne or direct-conversion receiver (DCR). DCR systems have advantages such as a simple architecture, easy integration with the baseband circuit and the elimination of the problem with the "image" signal in comparison with the heterodyne architecture.[18]

Mixers are inherently noisy circuits since noise contributions from multiple frequency bands are transferred to the output.[19] However, the requirement on the noise figure is relaxed for the stages following the LNA in a receiver system according to Friis formula (Eq. 2-4). Therefore, the noise figure of a mixer in a heterodyne system is not very critical (although still important) for the overall noise performance if an LNA with high enough gain is used.

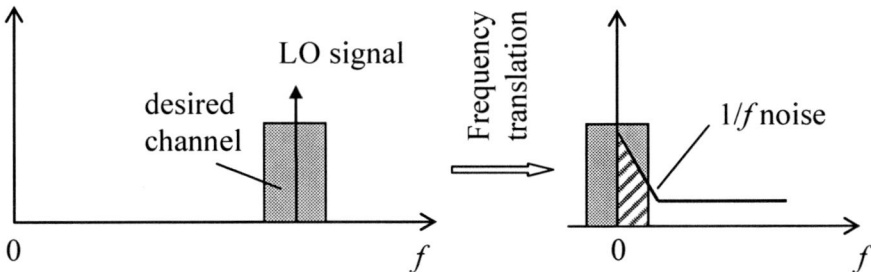

Figure 5-6. Frequency translation of signals performed in a mixer used in a DCR system.

For a DCR receiver, on the other hand, the $1/f$ noise in the transistors of the mixer can severely degrade the signal-to-noise ratio.[20,21] This can be understood because the signal is translated to very low frequencies where the $1/f$ noise of the devices contributes appreciable as illustrated in Fig. 5-6. The $1/f$ noise in the mixer is primarily generated in the switching transistors,[21] see Fig. 5-7. The $1/f$ noise generated in transistor on the RF input port is of minor significance for the total noise performance since the $1/f$ noise is upconverted to high frequencies (f_{LO}). The $1/f$ noise problem can be mitigated by employing long channel transistors in the mixer, which instead degrades the transconductance and the circuit speed. According to the model by Darabi and Abidi, sharper LO transitions (more square-like) as well as reduced DC bias current can decrease the noise.[21] In a novel approach, where the current is injected dynamically, lower $1/f$ corner was achieved.[22] The most common solution to the $1/f$ noise problem from a circuit point of view is to adopt a passive mixer, which involves no DC biasing current.[20,23] Passive mixers still exhibit $1/f$ noise at the output,[24] but the $1/f$ noise corner is reduced to a few tens of kHz. The drawback with passive mixers is that they provide no gain, which puts more stringent requirements on the noise performance of subsequent stages. Due to the lack of gain, passive mixers have been less preferred in RF design.

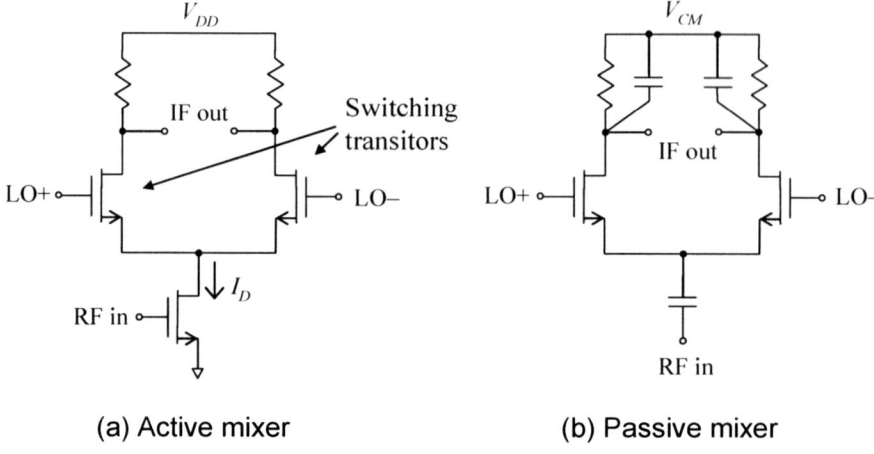

Figure 5-7. Single-balanced (a) active and (b) passive mixer topologies.

5. LOW-NOISE AMPLIFIER (LNA)

A low-noise amplifier is used in the first stage of a receiver system to amplify the weak signal received from the antenna. Designing the LNA with low internal noise is of enormous importance since the noise performance of the first circuit in a cascaded stage is most important (if the gain is >> 1). An RF amplifier is designed to work at frequencies in the GHz range, therefore only the thermal noise is important since the contribution from the device $1/f$ noise to the total noise power obviously is negligible. Upconversion of the $1/f$ noise to phase or amplitude modulated noise can take place due to non-linear effects,[25] but this is not a problem for CMOS LNAs. However, $1/f$ noise can be a problem for an amplifier working in a small bandwidth at low frequencies down to DC. The low-noise amplifiers typically used in LF noise measurement setups do show output $1/f$ noise. But in contrast to RF amplifiers, the speed of the devices employed in amplifiers working at low frequencies is not critical. The devices can therefore be designed to show minimum $1/f$ noise by making their size large.

The noise figure of an RF amplifier depends on the gain and the thermal noise of the transistor(s) as well as on the impedance matching network. The minimum noise figure of the CMOS transistors is reduced as their cut-off frequency increases for smaller gate lengths.[26,27] However, as scaling continues, increased gate leakage currents, greater impact of velocity saturation and increased importance of substrate resistance as well as higher $1/f$ noise due to miniaturization of device sizes will lead to higher noise.[3]

5. Introduction to noise in RF/analog circuits

SUMMARY

This chapter has presented an introduction to noise in mixers, VCOs and LNAs. The $1/f$ noise in MOS transistors is a problem for several analog and RF circuits, especially for VCOs and for mixers used in direct-conversion receiver (DCR) systems. The $1/f$ noise can be upconverted to undesired phase noise in a VCO. In a downconversion mixer used in a DCR system, the signals are translated to low frequencies where the $1/f$ noise is a severe problem. Due to the influence of $1/f$ noise in these circuits, the performance of a transceiver system used in communications can be limited, which imposes additional circuit design considerations in analog/RF design.

REFERENCES

1. B. Razavi, CMOS technology characterization for analog and RF design, *IEEE J. Solid-State Circuits* **34**, 268-276 (1999).
2. M. J. Deen and O. Marinov, Noise in advanced electronic devices and circuits, in *Proc. Int. Conf. Noise and Fluctuations (ICNF)*, 2005, pp. 3-12.
3. H. S. Bennett, R. Brederlow, J. C. Costa, P. E. Cottrell, W. M. Huang, A. A. Immorlica, J.-E. Mueller, M. Racanelli, H. Shichijo, C. E. Weitzel, and B. Zhao, Device and technology evolution for Si-based RF integrated circuits, *IEEE Trans. Electron Devices* **52**, 1235-1258 (2005).
4. J. Phillips and K. Kundert, Noise in mixers, oscillators, samplers and logic - An introduction to cyclostationary noise, in *Proc. IEEE Custom Integrated Circuits Conf.*, 2000, pp. 431-439.
5. B. Razavi, A study of phase noise in CMOS oscillators, *IEEE J. Solid-State Circuits* **31**, 331-343 (1996).
6. A. Hajimiri and T. H. Lee, A general theory of phase noise in electrical oscillators, *IEEE J. Solid-State Circuits* **33**, 179-194 (1998).
7. T. H. Lee and A. Hajimiri, Oscillator phase noise: a tutorial, *IEEE J. Solid-State Circuits* **35**, 326-336 (2000).
8. M. Sandén, F. Jonsson, O. Marinov, M. J. Deen, and M. Östling, Up-conversion of $1/f$ noise to phase noise in voltage controlled oscillators, in *Proc. 16th Int. Conf. Noise and Fluctuations (ICNF)*, 2001, pp. 499-502.
9. F. Jonsson, M. von Haartman, M. Sandén, M. Östling, and M. Ismail, A voltage controlled oscillator with automatic amplitude control in SiGe technology, in *Proc. 19th NORCHIP conf.*, 2001, pp. 28-33.
10. B. De Muer, M. Borremans, M. Steyaert, and G. Li Puma, A 2-GHz low-phase-noise integrated *LC*-VCO set with flicker-noise upconversion minimization, *IEEE J. Solid-State Circuits* **35**, 1034-1038 (2000).
11. S. Levantino, C. Samori, A. Bonfanti, S. L. J. Gierkink, A. L. Lacaita, and V. Boccuzzi, Frequency dependence on bias current in 5-GHz CMOS VCOs: impact on tuning range and flicker noise upconversion, *IEEE J. Solid-State Circuits* **37**, 1003-1011 (2002).
12. Y.-H. Kao and M.-T. Hsu, Theoretical analysis of low phase noise design of CMOS VCO, *IEEE Microwave and Wireless Components Lett.* **15**, 33-35 (2005).

13. A. A. Abidi, Phase noise and jitter in CMOS ring oscillators, *IEEE J. Solid-State Circuits* **41**, 1803-1816 (2006).
14. T. K. K. Tsang and M. N. El-Gamal, A high figure of merit and area-efficient low-voltage(0.7-1 V) 12 GHz CMOS VCO, in *Proc. IEEE Radio Frequency Integrated Circuits Symp.*, 2003, pp. 89-92.
15. A. Hajimiri and T. H. Lee, Design issues in CMOS differential *LC* oscillators, *IEEE J. Solid-State Circuits* **34**, 717-724 (1999).
16. H. Q. Liu, W. L. Goh, and L. Siek, 1.8-V 10-GHz ring VCO design using 0.18-μm CMOS technology, in *Proc IEEE Int. SOC Conf.*, 2005, pp. 77-78.
17. W.-H. Tu, J.-Y. Yeh, H.-C. Tsai, and C.-K. Wang, A 1.8V 2.5-5.2 GHz CMOS dual-input two-stage ring VCO, in *Proc. Asia-Pacific Conf. Advanced System Integrated Circuits*, 2004, pp. 134-137.
18. B. Razavi, *RF microelectronics* (Prentice-Hall, New Jersey, 1997).
19. M. T. Terrovitis and R. G. Meyer, Noise in current-commutating CMOS mixers, *IEEE J. Solid-State Circuits* **34**, 772-783 (1999).
20. S. Zhou and M.-C. F. Chung, A CMOS passive mixer with low flicker noise for low-power direct-conversion receiver, *IEEE J. Solid-State Circuits* **40**, 1084-1093 (2005).
21. H. Darabi and A. A. Abidi, Noise in RF-CMOS mixers: a simple physical model, *IEEE Trans. Solid State Circuits* **35**, 15-25 (2000).
22. H. Darabi and J. Chiu, A noise cancellation technique in active RF-CMOS mixers, *IEEE Int. Solid-State Circuits Conf.*, 2005, pp. 544-545.
23. T.-K. Nguyen, N.-J. Oh, V.-H. Le, and S.-G. Lee, A low-power CMOS direct conversion receiver with 3-dB NF and 30-kHz flicker-noise corner for 915-MHz band IEEE 802.15.4 ZigBee standard, *IEEE Trans. Microw. Theory Tech.* **54**, 735-741 (2006).
24. S. Chehrazi, R. Bagheri, and A. A. Abidi, Noise in passive FET mixers: a simple physical model, *IEEE Custom Integrated Circuits Conf.*, 2004, pp. 375-378.
25. T. D. Tomlin, K. Fynn, and A. Cantone, A model for phase noise generation in amplifiers, *IEEE Trans. Ultrason. Ferroelect. Freq. Contr.* **48**, 1547-1554 (2001).
26. K. Lee, I. Nam, I. Kwon, J. Gil, K. Han, S. Park, and B.-I. Seo, The impact of semiconductor technology scaling on CMOS RF and digital circuits for wireless application, *IEEE Trans. Electron Devices* **52**, 1415-1422 (2005).
27. P. H. Woerlee, M. J. Knitel, R. van Langevelde, D. B. M. Klaassen, L. F. Tiemeijer, A. J. Scholten, and A. T. A. Zegers-van Duijnhoven, RF-CMOS performance trends, *IEEE Trans. Electron Devices* **48**, 1776-1782 (2001).

PROBLEMS

1. Calculate the $1/f^3$-corner frequency for an oscillator with an impulse sensitivity function

$$ISF = 0.1 + \cos(\omega_0 t).$$

The drain current noise of the MOSFETs is given as

$$S_{I_D} = 1 \times 10^{-18} / f + 1 \times 10^{-23} \text{ A}^2/\text{Hz}.$$

2. An *LC* oscillator operating at 5.2 GHz is employed in a receiver. The frequency spectrum is divided in channels with 200-kHz bandwidth in the communication system the receiver is designed for. The worst-case interfering signal is specified to be at 2 MHz offset and 40 dB stronger than the desired channel. How low must the oscillator phase noise in dBc/Hz be at the frequency offset in question in order to achieve a SNR (due to phase noise) of at least 15 dB?

Hint: Study Fig. 5-2, which describes the same situation as in this problem.

Appendix I
LIST OF SYMBOLS

Table A-1. List of symbols.

Symbol	Unit	Meaning
A_{bulk}	-	BSIM parameter: bulk charge effect (see chapter 3.6)
af	-	SPICE noise parameter: current exponent (see chapter 3.6)
A_P	-	Available power gain of amplifier
C	F (F/cm^2)	Capacitance (per unit area)
C_{box}	F/cm^2	Capacitance per unit area of buried oxide (SOI)
C_d	F/cm^2	Depletion layer capacitance per unit area
C_{dm}	F/cm^2	Depletion layer capacitance per unit area when the depletion layer has its maximum width W_{dm}
C_{fb}	F/cm^2	Flat-band capacitance per unit area
C_{fox}	F/cm^2	Capacitance per unit area of front gate oxide (SOI)
C_G	F	Gate capacitance
C_{GS}	F	Gate-to-source capacitance
C_{it}	F/cm^2	Interface trap capacitance per unit area
C_{ox}	F/cm^2	Gate oxide capacitance per unit area
C_{Si}	F/cm^2	Capacitance per unit area of Si body (SOI)
d	m	Lattice constant
D_{it}	cm^{-2}eV^{-1}	Density of interface states (traps)
E	J, eV	Energy
E	V/cm	Electric field
E_C	eV	Conduction band edge energy
E_{eff}	V/cm	Effective electric field (effective vertical field in inversion layer)
ef	-	BSIM and SPICE noise parameter: frequency exponent (see 3.6)
E_F	eV	Fermi energy level

Symbol	Unit	Meaning
$E_{F,n}$	eV	Quasi-Fermi energy level for electrons
$E_{F,p}$	eV	Quasi-Fermi energy level for holes
E_g	eV	Band gap energy
E_i	eV	Intrinsic Fermi energy level
E_T	eV	Trap energy level
E_V	eV	Valence band edge energy
F	-	Noise factor
f	Hz	Frequency
$f(E)$	-	Fermi-Dirac distribution function, gives the probability that an electronic state at energy E is occupied
$f(X)$	-	Probability density function of random variable X
FF	-	Fano factor
f_T	Hz	Transition (cut-off) frequency
f_0	Hz	Oscillation frequency of VCO
g	-	Degeneracy factor
$g(\tau)$		Trap distribution function (chapter 1.3.5)
g_{ch}	S = A/V	Channel conductance
$g_{ch,0}$	S	Channel conductance at zero drain-source voltage
g_m	S	Transconductance
h	Js	Planck's constant ($= 6.63 \times 10^{-34}$ Js)
I	A	Current
i	A	Small-signal current
I_D	A	Drain current
$I_{D,sat}$	A	Drain current in saturation
I_G	A	Gate leakage current
i_n	A	Noise current
$\overline{i_n^2}$	A^2	Mean square of noise current
I_S	A	Source current
I_0	A	Diode saturation current
J	A/cm^2	Current density
k	J/K	Boltzmann's constant ($= 1.38 \times 10^{-23}$ J/K)
k	-	Dielectric constant (relative permittivity)
KF		SPICE noise parameter (see chapter 3.6)
L	cm	Gate length (length)
$L(\Delta f)$	dBc/Hz	Phase noise at offset Δf from carrier
L_{eff}	m	BSIM parameter: effective gate length (see chapter 3.6)
m	-	MOSFET body-effect coefficient
M	-	Avalanche multiplication factor
m^*	kg	Electron (hole) effective mass
N	-	Number of carriers

Appendix I

Symbol	Unit	Meaning
N^*	m^{-2}	BSIM noise parameter (see chapter 3.6)
n	cm^{-3}	Electron concentration (per unit volume)
N_a	cm^{-3}	Acceptor doping concentration
N_{body}	cm^{-3}	Doping concentration in Si body of SOI substrate
N_d	cm^{-3}	Donor doping concentration
NF	dB	Noise figure
n_i	cm^{-3}	Intrinsic carrier concentration (= 1.5×10^{10} cm^{-3} for Si at 300K)
N_l	m^{-2}	BSIM parameter: charge density at drain side (see chapter 3.6)
$NOIA$	m^{-3}eV^{-1}	BSIM noise parameter (see chapter 3.6)
$NOIB$	m^{-1}eV^{-1}	BSIM noise parameter (see chapter 3.6)
$NOIC$	m·eV^{-1}	BSIM noise parameter (see chapter 3.6)
N_{ox}	cm^{-2}	Oxide charge density (= Q_{ox}/q)
N_{OX}	-	Number of oxide charges
n_s	cm^{-3}	Surface carrier concentration
N_s	cm^{-2}	Channel carrier density (= Q_i/q)
N_{sub}	cm^{-3}	Doping concentration in the substrate
N_t	cm^{-3}eV^{-1}	Oxide trap density (per unit volume)
N_T		Number of oxide traps
N_0	m^{-2}	BSIM parameter: charge density at source side (see chapter 3.6)
p	cm^{-3}	Hole concentration (per unit volume)
P	W	Power
P	-	Probability (chapter 1.2.2)
P_n	W	Available noise power
q	C	Electronic charge (= 1.602×10^{-19} C)
Q_d	C/cm^2	Depletion charge per unit area
Q_i	C/cm^2	Inversion charge per unit area
Q_m	C/cm^2	Charge on gate per unit area
q_{max}	C	Maximum charge displacement of tank capacitor in VCO
Q_{ox}	C/cm^2	Oxide charge per unit area
R	Ω	Resistance
$R(s)$		Autocorrelation function (chapter 1.2.2)
r_{ch}	Ω	Channel resistance
R_D	Ω	Drain series resistance
R_{in}	Ω	Input resistance
R_L	Ω	Load resistance
R_n	Ω	Noise resistance (Eq. 1-12)
r_n	Ω	Equivalent noise resistance (Eq. 2-7)
R_S	Ω	Source series resistance
R_{SD}	Ω	Source-drain series resistance ($R_S + R_D$)
r_π	Ω	Dynamic resistance for pn-junction
S		Power spectral density

Appendix I

Symbol	Unit	Meaning
S_I	A²/Hz	Power spectral density of current fluctuations
S_{I_D}	A²/Hz	Power spectral density of drain current noise
$S_{I_{D,ch}}$	A²/Hz	Power spectral density of drain current noise in the channel
$S_{I_{D,th}}$	A²/Hz	Power spectral density of drain current thermal noise
$S_{I_{D,1/f}}$	A²/Hz	Power spectral density of drain current $1/f$ noise
S_{I_G}	A²/Hz	Power spectral density of gate current noise
$S_{I_{R_D}}$	A²/Hz	Power spectral density of current noise generated in R_D
$S_{I_{R_S}}$	A²/Hz	Power spectral density of current noise generated in R_S
$S_{I_{R_{SD}}}$	A²/Hz	Power spectral density of current noise generated in R_{SD}
$S_{\lim}$	A²/Hz	BSIM noise parameter (see chapter 3.6)
S_N	1/Hz	Power spectral density of carrier number fluctuations
$S_{Q_{ox}}$	C²/cm⁴Hz	Power spectral density of oxide charge density fluctuations
S_R	Ω²/Hz	Power spectral density of resistance fluctuations
S_V	V²/Hz	Power spectral density of voltage fluctuations
$S_{V_{fb}}$	V²/Hz	Power spectral density of flat-band voltage noise
S_{V_G}	V²/Hz	Power spectral density of equivalent input gate voltage noise
S_{wi}	A²/Hz	BSIM noise parameter (see chapter 3.6)
SS	V/decade	Subthreshold slope
T	K	Absolute temperature
T	s	Time (constant)
t	s	Time
t_{box}	cm	Thickness of buried oxide (SOI)
t_{EOT}	cm	Equivalent oxide thickness
T_n	K	Noise temperature (Eq. 1-11)
t_{ox}	cm	Gate oxide thickness
t_{Si}	cm	Thickness of Si body in SOI substrate
T_0	K	Standard noise temperature (= 290 K)
V	cm³	Volume
V	V	Voltage
v	V	Small-signal voltage
V_B	V	Substrate voltage (Bulk terminal voltage)
V_{BS}	V	Substrate-to-source voltage
v_d	cm/s	Carrier drift velocity
V_d	V	Applied voltage across pn-junction
V_{DD}	V	Power supply voltage
V_{DS}	V	Drain-to-source voltage
$V_{DS,sat}$	V	MOSFET drain-to-source saturation voltage
V_{fb}	V	Flat-band voltage
V_G	V	Gate voltage

Appendix I

Symbol	Unit	Meaning
V_{GS}	V	Gate-to-source voltage
V_{GT}	V	Gate voltage overdrive ($= |V_{GS} - V_T|$)
v_i	cm/s	Individual carrier drift velocity
V_{max}	V	Maximum voltage swing over tank capacitor in VCO
v_n	V	Noise voltage
$\overline{v_n^2}$	V²	Mean square of noise voltage
$v_{n,rms}$	V	RMS noise voltage
V_T	V	Threshold voltage
v_{th}	cm/s	Thermal velocity of electrons
W	cm	Gate width
W_d	cm	Depletion layer width
W_{dm}	cm	Maximum depletion layer width
W_{eff}	m	BSIM parameter: effective gate width (see chapter 3.6)
Y_S	S	Source admittance
z	cm	Distance in a direction vertical to the channel
Z_S	Ω	Source impedance
z_t	cm	Trap distance from gate oxide/channel interface
α	Vs/C	Scattering parameter of the correlated mobility fluctuations
α_C	Vs/C	Coulomb scattering parameter
α_H	-	Hooge parameter
$\alpha_{H,a}$	-	Hooge parameter of $1/f$ noise generated in scattering processes other than surface roughness scattering.
$\alpha_{H,ph}$	-	Hooge parameter of $1/f$ noise generated in the phonon scattering
$\alpha_{H,sr}$	-	Hooge parameter of $1/f$ noise generated in the surface roughness scattering
δ	m	Skin depth (Eq. 2-1)
Δf	Hz	Frequency separation from the oscillating frequency of a VCO
ΔL_{clm}	m	BSIM parameter: channel length reduction (see chapter 3.6)
Δx or δx		Fluctuation in x
ε_{ox}	F/cm	Permittivity of SiO_2 ($= 3.45 \times 10^{-13}$ F/cm)
ε_{Si}	F/cm	Silicon Permittivity ($= 1.04 \times 10^{-12}$ F/cm)
Φ_B	J, eV	Energy barrier height
ϕ_{ms}	V	Work-function difference between the gate material and the substrate material
γ	-	Frequency exponent
γ	-	MOSFET thermal noise coefficient
η	-	Electric field parameter (in Eq. 3-66)
η_v	-	MOSFET parameter describing the relative degree of drain saturation (in Eqs. 3-21 and 3-22)
κ_D	-	Gate leakage current partitioning coefficient at drain side

Symbol	Unit	Meaning
κ_S	-	Gate leakage current partitioning coefficient at source side
λ	cm	Tunneling attenuation length
λ_e	m	Phonon mean free path
μ	cm^2/Vs	Carrier mobility
μ_a	cm^2/Vs	Carrier mobility limited by other mechanisms than surface roughness scattering
μ_{ac}	cm^2/Vs	Mobility limited by scattering with surface acoustic phonons
μ_b	cm^2/Vs	Mobility limited by scattering with bulk phonons
μ_C	cm^2/Vs	Mobility limited by Coulomb scattering
$\mu_{C,imp}$	cm^2/Vs	Mobility limited by Coulomb scattering from impurities
$\mu_{C,ox}$	cm^2/Vs	Mobility limited by Coulomb scattering from oxide charges
μ_{C0}	cm/Vs	Screened Coulomb scattering parameter
μ_{eff}	cm^2/Vs	Effective mobility in MOSFET inversion layer
μ_i	cm^2/Vs	Individual carrier mobility
μ_{ph}	cm^2/Vs	Mobility limited by scattering with phonons, both bulk phonons and surface acoustic phonons.
μ_r		Relative permeability (in Eq. 2-1)
μ_{sr}	cm^2/Vs	Mobility limited by surface roughness scattering
μ_0	cm^2/Vs	Low-field mobility (Eq. 3-78)
μ_0	H/m	Permeability of free space (= $4\pi \times 10^{-7}$ H/m), used in Eq. (2-1)
θ	rad	Phase
θ	V^{-1}	Mobility attenuation coefficient (in Eq. 3-78)
$\rho_{1,2}$	-	Correlation coefficient
σ	Ω^{-1}cm^{-1}	Conductivity
σ		Standard deviation (in Eq. 1-5)
σ_e	cm^2	Capture cross section for electrons
σ_h	cm^2	Capture cross section for holes
σ_{N_t}	cm$^{0.5}$eV$^{0.5}$	Relative standard deviation of the trap density
τ	s	CMOS inverter delay
τ	s	Time constant of g-r noise
τ_c	s	Capture time for electrons (holes)
τ_e	s	Emission time for electrons (holes)
τ_h	s	Time in high level of two-state RTS
τ_l	s	Time in low level of two-state RTS
τ_{ph-ph}	s	Relaxation time for phonon-phonon scattering
τ_{th}	s	Time constant of thermally activated traps
τ_0	s	Tunneling time constant (usually taken as 10^{-10} s)
ω_0	rad/s	Angular frequency of oscillation
ψ_B	V	Difference between Fermi level and intrinsic level potentials
ψ_s	V	Surface potential

Appendix I

1 eV = 1.602×10^{-19} J
1 Å = 10^{-10} m
log means logarithm function in base 10
ln means natural logarithm (base e)

Appendix II
LIST OF ACRONYMS

Table A-2. List of Acronyms.

Acronym	Meaning
ALD	Atomic layer deposition
B	Bulk
BSIM	Berkeley short channel IGFET model
CMOS	Complementary metal-oxide-semiconductor
CVD	Chemical vapour deposition
D	Drain
DC	Direct current
DCR	Direct conversion receiver
DIBL	Drain-induced barrier lowering
DT	Dynamic threshold
DUT	Device-under-test
EOT	Equivalent oxide thickness
FD	Fully depleted
FFT	Fast Fourier Transform
FOM	Figure-of-merit
FUSI	Fully silicided
G	Gate
g-r	Generation-recombination
HF-clean	Clean in Hydrofluoric acid
ISF	Impulse sensitivity function
ITRS	International technology roadmap of semiconductors
I-V	Current-voltage
JFET	Junction field-effect transistor

Acronym	Meaning
LC	Inductance-capacitance
LNA	Low-noise amplifier
MBE	Molecular beam epitaxy
MOCVD	Metal-organic chemical vapour deposition
MOSFET	Metal-oxide-semiconductor field-effect transistor
nMOSFET	n-channel MOS transistor
PD	Partially depleted
pMOSFET	p-channel MOS transistor
PSD	Power spectral density
PVD	Physical vapour deposition
RF	Radio frequency
RTS	Random-telegraph-signal
S	Source
SCE	Short channel effect
SNR	Signal-to-noise ratio
SOI	Silicon-on-insulator
TCAD	Technology computer aided design
TEM	Transmission electron microscopy
VCO	Voltage controlled oscillator
WKB	Wentzel-Kramers-Brillouin

Appendix III
SOLUTIONS TO PROBLEMS

Solution problem 1-1.

We calculate the mean square noise voltage for the different alternatives.

A: Only thermal noise

$$\overline{v_n^2} = 4kTR\Delta f = 4\times 1.38\times 10^{-23}\times 300\times 5000\times 10^6 \text{ V}^2 = 8.28\times 10^{-11} \text{ V}^2.$$

B: $T_n = 500$ K

$$\overline{v_n^2} = 4kT_n R\Delta f = 1.38\times 10^{-10} \text{ V}^2.$$

C: $v_{n,rms} = 15$ μV

$$\overline{v_n^2} = v_{n,rms}^2 = 2.25\times 10^{-10} \text{ V}^2.$$

D: Power spectral density $S_V = 2\times 10^{-16}$ V^2/Hz

$$\overline{v_n^2} = S_V \times \Delta f = 2\times 10^{-16}\times 10^6 \text{ V}^2 = 2\times 10^{-10} \text{ V}^2.$$

E: $R_n = 10$ kΩ

$$\overline{v_n^2} = 4kTR_n\Delta f = 1.656\times 10^{-10} \text{ V}^2.$$

Answer: A, B, E, D, C.

Solution problem 1-2

The thermal noise from the resistor is

$$S_{I,th} = 4kT/R = 4\times 1.38\times 10^{-23} \times 300/200 \text{ A}^2/\text{Hz} = 8.28\times 10^{-23} \text{ A}^2/\text{Hz}.$$

The thermal noise and the $1/f$ noise are uncorrelated. Thus

$$S_{I,tot}(f) = 8.28\times 10^{-23} + 2.5\times 10^{-19}/f \quad \text{A}^2/\text{Hz}.$$

The noise power is given by integrating the PSD over the bandwidth

$$\overline{i_n^2} = \int_1^{10^4} S_{I,tot} df = 8.28\times 10^{-23} \times (10^4 - 1) + 2.5\times 10^{-19} \times \ln\left(\frac{10^4}{1}\right) \text{ A}^2 =$$

$$8.3\times 10^{-19} + 2.3\times 10^{-18} \text{ A}^2 = 3.1\times 10^{-18} \text{ A}^2 \Rightarrow i_{n,rms} = \sqrt{\overline{i_n^2}} = 1.8 \text{ nA}.$$

Solution problem 1-3

The noise equivalent circuit is shown in Fig. A-1.

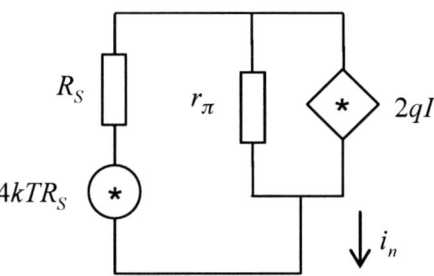

Figure A-1. Noise equivalent circuit in example 1-3. $r_\pi = kT/qI$.

The two noise sources are uncorrelated. We use the superposition principle to calculate the noise current from each source.

From the thermal noise source using Ohm's law:

$$S_{I_n,1} = \frac{4kTR_S}{(R_S + r_\pi)^2}.$$

From the shot noise source using the current-division principle

$$S_{I_n,2} = 2qI \frac{r_\pi^2}{(R_S + r_\pi)^2}.$$

Answer: The total PSD is according to the superposition principle

$$S_{I_n} = S_{I_n,1} + S_{I_n,2}.$$

Solution problem 1-4

The measured noise consists of superimposed $1/f$ noise and white noise (thermal noise). From the graph:

$$S_V = 2 \times 10^{-13} / f + 5 \times 10^{-16} \text{ V}^2/\text{Hz}.$$

The resistance can be calculated from the thermal noise level

$$R = \frac{5 \times 10^{-16}}{4kT} \ \Omega = 30 \text{ k}\Omega.$$

The Hooge noise model (see Eq. 1-37)

$$\frac{S_R}{R^2} = \frac{S_V}{V^2} = \frac{\alpha_H}{fN} \quad (V = RI \Rightarrow S_V = S_R I^2 \Rightarrow S_V / V^2 = S_R / R^2).$$

We need to determine the number of (free) electrons N
N = electron concentration × sample volume $\Rightarrow$

$$N = 10^{17} \text{ cm}^{-3} \times 10^{-3} \text{ cm} \times 10^{-2} \text{ cm} \times 10^{-4} \text{ cm} = 10^8.$$

Answer: $\alpha_H = \dfrac{fNS_V}{V^2} = \dfrac{10^8 \times 2 \times 10^{-13}}{(30 \times 10^3 \times 16.6 \times 10^{-6})^2} = 8 \times 10^{-5}$.

Solution problem 2-1

The noise equivalent circuit is shown in Fig. A-2.

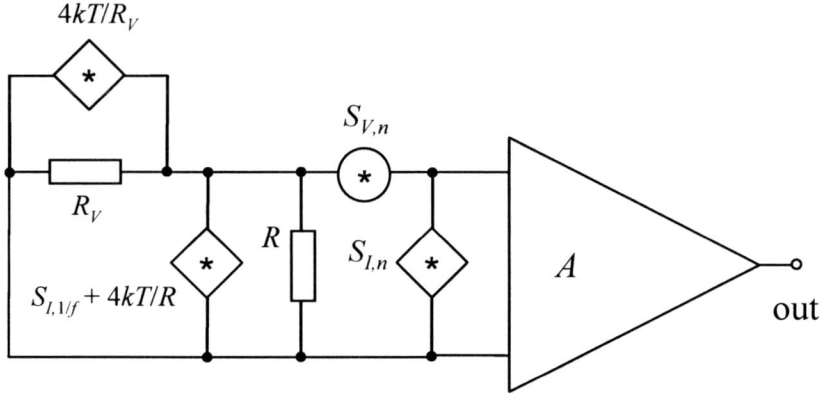

Figure A-2. Noise equivalent circuit in example 2-1.

Answer:
The voltage noise at the output can be written according to the superposition principle:

$$S_{V,out} = \left(\dfrac{4kT/R_V \cdot R_V^2 R^2}{(R+R_V)^2} + \dfrac{(S_{I,1/f} + 4kT/R) \cdot R_V^2 R^2}{(R+R_V)^2} \right.$$

$$\left. + S_{V,n} + \dfrac{S_{I,n} \cdot R_V^2 R^2}{(R+R_V)^2} \right) \cdot A^2$$

$$= \left(\dfrac{4kTR_V R}{(R+R_V)} + \dfrac{S_{I,1/f} R_V^2 R^2}{(R+R_V)^2} + S_{V,n} + \dfrac{S_{I,n} \cdot R_V^2 R^2}{(R+R_V)^2} \right) \cdot A^2.$$

Solution problem 2-2

Using Eqs. (2-8) to (2-10), $p = 2$ for number fluctuation noise and

Appendix III

$g_{ch}(R_{SD} + R_L) \ll 1$ gives

$$S_{I_{D,tot}} = \frac{A \cdot I_D^2}{fWL(V_{GS} - V_T)^2} + g_{ch}^2 R_{SD}^2 \frac{B \cdot I_D^2}{f}$$

where A and B are constants. At a constant V_{DS} in the linear regime

$$g_{ch} = \left(\frac{dI_D}{dV_{DS}}\right) = \left\{I_D = \frac{W}{L}\mu C_{ox}(V_{GS} - V_T)V_{DS}\right\} = I_D/V_{DS}.$$

Thus,

$$S_{I_{D,tot}} = \frac{A \cdot I_D^2}{fWL(V_{GS} - V_T)^2} + \frac{R_{SD}^2}{V_{DS}^2} \frac{B \cdot I_D^4}{f}.$$

It was given that the noise contributions from the S/D resistance and the channel are equally strong at $I_D = I_{D,1}$. Thus,

$$\frac{A \cdot I_{D,1}^2}{fWL(V_{GS,1} - V_T)^2} = \frac{R_{SD}^2}{V_{DS}^2} \frac{B \cdot I_{D,1}^4}{f} \Rightarrow$$

$$\frac{R_{SD}^2}{V_{DS}^2} \frac{B}{f} = \frac{A \cdot I_{D,1}^{-2}}{fWL(V_{GS,1} - V_T)^2}.$$

At $I_D = I_{D,1}/3$:

$$I_D \propto (V_{GS} - V_T) \Rightarrow (V_{GS} - V_T) = (V_{GS,1} - V_T)/3.$$

The relative contribution from the 1/f noise in the S/D resistance to the output drain current noise equals

$$\frac{\dfrac{R_{SD}^2}{V_{DS}^2}\dfrac{B\cdot I_D^4}{f}}{\dfrac{R_{SD}^2}{V_{DS}^2}\dfrac{B\cdot I_D^4}{f}+\dfrac{A\cdot I_D^2}{fWL(V_{GS}-V_T)^2}} = \frac{1}{1+\dfrac{A\cdot I_{D,1}^2/3^2}{fWL(V_{GS,1}-V_T)^2/3^2}\cdot\dfrac{fWL(V_{GS,1}-V_T)^2}{A\cdot I_{D,1}^{-2}\cdot I_{D,1}^4/3^4}}$$

$$= \frac{1}{1+3^4} \approx 1.2\%.$$

Answer: The noise from the S/D resistance is only 1.2% of the total output drain current noise at $I_D = I_{D,1}/3$.

Solution problem 2-3

RTS noise can be observed on top of the $1/f$ noise when the RTS noise PSD is larger than the $1/f$ noise PSD.

Using Eq. (1-37) for the $1/f$ noise

$$S_{I,1/f} = \frac{\alpha_H \cdot I_D^2}{fN}$$

and Eq. (1-31) for the RTS noise with $N_T = 1$ (one trap)

$$S_{I,RTS} = \frac{I_D^2}{N^2}\frac{\tau}{1+(2\pi f)^2\tau^2}.$$

The RTS noise PSD is maximized relative to the $1/f$ noise when $\tau = 1/2\pi f$.

Answer: the condition to observe RTS noise

$$S_{I,RTS} > S_{I,1/f} \Rightarrow$$

$$\frac{I_D^2}{N^2}\frac{\tau}{2} > \frac{\alpha_H I_D^2 2\pi\tau}{N}$$

$$N < \frac{1}{4\pi\alpha_H}.$$

Appendix III

Solution problem 2-4

RTS noise can be observed if
(i) the number of traps is small *and*
(ii) the RTS noise is higher than the $1/f$ noise.

We make the assumption that RTS can be observed if there are less than 5 active traps.
Thus, using Eq. (2-12)

$$4kTWLN_t z < 5 \Rightarrow WL < \frac{5}{4kTN_t z}.$$

It was given that $z = 2$ nm and $N_t = 1\times10^{17}$ cm^{-3}eV^{-1}. Hence,

$$WL < 0.24 \; \mu\text{m}^2.$$

The number of carriers N must also satisfy Eq. (2-11). For a MOSFET in inversion

$$N = WLC_{ox}(V_{GS} - V_T)/q$$

Thus,

$$N < \frac{1}{4\pi\alpha_H} \Rightarrow \alpha_H < \frac{q}{4\pi WLC_{ox}(V_{GS} - V_T)}.$$

This relation is valid in strong inversion. The RTS noise is easiest to discover when $V_{GS} - V_T$ is small. Therefore, we assume $V_{GS} - V_T = 0.1$ V. Furthermore, $C_{ox} = \varepsilon_{ox}/t_{ox} = 1.73\cdot10^{-6}$ F/cm^2. Thus, the requirement on α_H is found to be

$$\alpha_H < 3\cdot10^{-5}.$$

<u>Answer:</u> RTS noise is estimated to be observed in a device with a gate area $W \cdot L$ smaller than 0.24 μm^2 and with a Hooge parameter for background $1/f$ noise lower than $3\cdot10^{-5}$.

Solution problem 3-1

Insert $V_{DS,sat} = (V_{GS} - V_T)/m$ in Eq. (3-50):

$$\frac{S_{I_D}}{I_D^2} = \frac{q\alpha_H \mu_{eff} V_{DS}}{fL^2 I_D} = \frac{q\alpha_H \mu_{eff}(V_{GS} - V_T)}{mfL^2 I_D}. \quad \text{(A3-1)}$$

Then insert Eq. (3-8) for I_D in the equation above

$$I_D = \frac{W}{L}\mu_{eff} C_{ox} \frac{(V_{GS} - V_T)^2}{2m} \Rightarrow$$

$$S_{I_D} = \frac{2q\alpha_H I_D^2}{fWLC_{ox}(V_{GS} - V_T)}. \quad \text{(A3-2)}$$

Comparing with Eq. (3-80) for mobility fluctuation noise in the linear region, we see that the drain current noise is a factor of two higher in the saturation region. Why? The number of carriers in the channel is reduced to 2/3 of its value in the linear region at a certain gate voltage overdrive since the channel at the drain end is pinched off. By using

$$Q_i = C_{ox}(V_{GS} - V_T)\sqrt{1 - x/L}. \quad \text{(A3-3)}$$

in Eq. (3-50), the same expression is as in Eq. (A3-2) is found.

Solution problem 3-2

First, calculate the expression for g_m in the saturation region from Eq. (3-8)

$$g_{m,sat} = \left(\frac{dI_{D,sat}}{dV_{GS}}\right) = \frac{2I_D}{(V_{GS} - V_T)} = \sqrt{2I_D \frac{W}{L} \frac{\mu_{eff} C_{ox}}{m}}. \quad \text{(A3-4)}$$

Inserted in Eq. (3-79), this yields

$$S_{I_D} = \frac{q^2 kT\lambda N_t}{f^\gamma WLC_{ox}^2}\left(1 + \frac{\alpha\mu_{eff}C_{ox}I_D}{g_m}\right)^2 g_m^2 =$$

$$\frac{q^2 kT\lambda N_t}{f^\gamma WLC_{ox}^2}\left(1 + \frac{\alpha\mu_{eff}C_{ox}(V_{GS}-V_T)}{2}\right)^2 \frac{2I_D W}{L}\frac{\mu_{eff}C_{ox}}{m} = \quad\quad (A3\text{-}5)$$

$$\frac{2q^2 kT\mu_{eff}I_D\lambda N_t}{f^\gamma L^2 C_{ox} m}\left(1 + \frac{\alpha\mu_{eff}C_{ox}(V_{GS}-V_T)}{2}\right)^2.$$

This expression has the same functional dependence as Eqs. (3-68). The BSIM3 noise model is based on the number fluctuation noise theory. The SPICE2 model in Eq. (3-67) resembles the pure number fluctuation noise model (without correlated mobility fluctuations) in the saturation region.

Solution problem 3-3

(a) Mobility fluctuations in the linear or subthreshold region (see Eq. 3-80), or number fluctuations in the saturation region (see Eq. A3-5).
(b) Mobility fluctuations in the saturation region. See Eq. (3-51).
(c) Number fluctuations in the linear region. See Eq. (3-39) and note that $V_{GS} - V_T \propto I_D$.
(d) The output noise stems from noise in the drain series resistance. See Eq. (3-12) with

$$S_{I_{R_D}} \gg S_{I_{D,ch}} \text{ and } r_{ch} > R_D.$$

Fig. 2-10 shows a simulation of the situation described above.
(e) Two alternatives: the noise stems (i) from number fluctuation noise in the channel when biased in the subthreshold region (see Eq. 3-40) or (ii) from noise in the drain series resistance when

$$S_{I_{R_D}} > S_{I_{D,ch}} \text{ and } R_D > r_{ch}.$$

Solution problem 4-1

First calculate the oxide capacitance C_{ox} and the transconductance g_m.

$$C_{ox} = 3.9 \cdot 8.854 \cdot 10^{-14}/(2.2 \cdot 10^{-7}) \text{ F/cm}^2 = 1.57 \cdot 10^{-6} \text{ F/cm}^2.$$

Use Eq. (3-8) to find g_m.

$$g_m = \frac{I_D}{[(V_{GS}-V_T)-mV_{DS}/2]} = 1.30 \text{ mS if we assume } m=1.$$

(a) <u>Answer</u>: Eq. (3-80) $\Rightarrow \alpha_H = 7.9\times10^{-5}$.

The trap density is found from Eq. (3-79). The correlated mobility fluctuations can be neglected since the gate voltage overdrive is small (0.1 V). Thus,

$$N_t = \frac{10\cdot10\cdot10^{-4}\cdot0.075\cdot10^{-4}\cdot(1.57\cdot10^{-6})^2}{1.602\cdot10^{-19}\cdot1.38\cdot10^{-23}\cdot300\cdot10^{-8}\cdot(0.0013)^2}\cdot1.4\cdot10^{-17}$$
cm^{-3}eV^{-1}.

Here, $\lambda = 10^{-8}$ cm was used (see Table 4-4).

<u>Answer</u>: $N_t = 2.3\times10^{17}$ cm^{-3}eV^{-1}.
(note that in order to get the unit eV, multiplication with 1.602×10^{-19} is performed).

(b) Calculate the input gate voltage noise and normalize with gate area

$$WLS_{V_G} = WLS_{I_D}/g_m^2 = 6.2 \text{ μV}^2\text{μm}^2/\text{Hz (at } f=10 \text{ Hz)}.$$

According to Table 4-1, the ITRS requires a noise level below 19 μV^{2}μm^2/Hz at $f = 10$ Hz for a device with $L = 75$ nm and $t_{EOT} = 2.2$ nm (note that the value is given at 1 Hz in the table).

<u>Answer</u>: The ITRS requirements are fulfilled.

Solution to problem 4-2

The input gate voltage noise is calculated in comparison with the reference device for all four technologies.

<u>Answer</u>:

i) $S_{V_G} = S_{V_G,ref} \dfrac{N_t/N_{t,ref} \cdot (t_{EOT}/t_{EOT,ref})^2}{L/L_{ref}} = S_{V_G,ref} \cdot \dfrac{6 \cdot (1/2)^2}{(1/2)} = 3S_{V_G,ref}$.

ii) Mobility fluctuation noise: $S_{V_G} = S_{V_G,ref} \cdot \dfrac{2 \cdot (3/4)}{(3/4)} = 2S_{V_G,ref}$.

iii) A front-back gate coupling factor $(1+N_{t,b}/N_{t,f})$ must be considered in the FD SOI device (see Eq. 4-16).

$$S_{V_G} = S_{V_G,ref} \cdot \dfrac{1}{(2/3)}(1+3) = 6S_{V_G,ref}.$$

iv) $S_{V_G} = S_{V_G,ref} \cdot 2^2 = 4S_{V_G,ref}$.

Solution to problem 5-1

Identifying Fourier coefficients: (see Eq. 5-1)
$c_0 = 0.2$, $c_1 = 1$.

The $1/f^3$-corner frequency is the frequency Δf where the first term and the second term in Eq. (5-2) are equal. Thus

$$0.2^2 \cdot 1 \cdot 10^{-18} / \Delta f = 1 \cdot 10^{-23}(0.2^2 + 1^2) \Rightarrow \Delta f = 3.8 \text{ kHz}.$$

Answer: The $1/f^3$-corner frequency is equal to 3.8 kHz.

Solution to problem 5-2

Denote the noise power (in dBm) due to phase noise with P_N, the power of the interfering signal P_I and the power of the signal in the desired channel P_C. Thus, in order to achieve a SNR of 15 dB

$$\text{SNR} = P_C - P_N = P_I - 40 \text{ dB} - P_N = 15 \text{ dB}. \tag{A5-1}$$

The phase noise with respect to the power of the interfering signal is denoted by L (dBc/Hz). The noise power is integrated over the channel bandwidth of 200 kHz

$$P_N = L + 10\log(200 \cdot 10^3) + P_I. \tag{A5-2}$$

Inserted in Eq. (A5-1), this yields

$$15 = -40 - L - 10\log(200 \cdot 10^3) \Rightarrow L = -40 - 53 - 15 = -108 \text{ dBc/Hz}.$$

<u>Answer:</u> The phase noise at 2 MHz offset must be lower than −108 dBc/Hz.

INDEX

$1/f$ noise
   general description 14
   other *See* under low-frequency noise

accumulation mode (AM) SOI 130, 133
Al_2O_3 47, 142, 145, 156
amplifier noise model 31
amplitude modulation noise 176
analog circuits 53, 128, 175
annealing 74, 83, 87, 149, 160
atomic layer deposition (ALD) 139
autocorrelation function 3, 4
available noise power 8
available power gain 32
avalanche multiplication 63

ballistic transport 108, 112
bandgap 11, 114
barrier height 68, 144
baseband 178, 182
battery 39
Berkeley short channel IGFET (BSIM3) model 92
bias circuit 30
bipolar transistors 13, 106
Boltzmann's constant 6
buried channel 45, 78, 84, 109, 115, 117, 130, 160
buried oxide 132
burst noise *See* random-telegraph-signal (RTS) noise

cables 39
calibration 39
capture cross section 46
capture time 46
carrier 7, 9, 59
carrier mobility 7
carrier scattering 59, 81
cascaded amplifiers 32

central limit theorem 3
channel length modulation 57, 63
channel orientation 127
charge-pumping measurements 41, 48, 126, 142
charge-sharing model 105
circuit simulations 92
cleaning 83, 109, 147
CMOS technology
   $1/f$ noise performance 109
   downscaling 103, 104, 109, 184
   general 53, 103
   Ge-on-insulator 123
   high-k gate dielectrics 48, 108, 137, 157
   metal gate 108, 157
   multiple-gate devices 108, 130, 159
   non-classical 106
   SiGe pMOSFETs 84, 114, 134, 151
   SiGe-on-insulator 123
   silicon-on-insulator (SOI) 58, 84, 108, 116, 127
   source/drain engineering 108
   strained Si 107, 123
coherence measurements 40
collision time 59
computer controlled measurements 31
conduction band 47, 70, 76, 124
conductivity 15
contacts 39, 45, 136
correlated mobility fluctuations 155
correlation 19, 76, 150
Coulomb scattering 59, 71, 112, 140
Coulomb scattering parameter 72
coupling effect 132
crystal direction 159
crystalline quality 15, 82, 83
current crowding 13
current fluctuations 7
cyclostationary noise 176

decibel (dB) scale 8
density of interface states 122
detrapping 13, 68
device architecture 109
diagnostic tool 41
dielectric constant 138
digital circuits 175
digital signal processing 34
diode 9
diode current 9
direct-conversion receiver 182
discrete levels 12
dislocations 125, 127
disturbances 2, 28, 39
doping concentration 47, 56, 106
double-gate SOI MOSFETs 130
downscaling *See* under CMOS technology
drain 54
drain current fluctuations 65
drain-induced-barrier-lowering (DIBL) 105
drift velocity 7
dynamic resistance 10
dynamic threshold (DT) MOSFETs 88

effective electric field 59, 82, 87, 88, 91, 122
effective mass, carrier 59, 76, 115, 124, 144
electric field 106
electron charge 7
electrons 7, 11, 59, 73
emission time 46
equivalent noise generators 31
ergodic process 4

Fano factor 10
Fast Fourier Transform (FFT) 34
Fermi level 11, 13, 45, 47, 56, 70, 118
Fermi-Dirac distribution 67
figure-of-merit, VCO 181
FinFETs 130, 159
fixed charge 59
flattop 35
flicker noise *See* 1/*f* noise
floating body effects 129, 131
frequency domain 34, 36
frequency exponent 68

frequency span 35
frequency spectrum 14
Friis formula 32, 182
fully depleted (FD) SOI 129, 132
fully silicided (FUSI) 157

GaAs 151
gain 29, 104, 175, 184
gate 54
gate oxidation 45, 83, 109
gate-all-around MOSFETs 130, 159
Gaussian top 35
Ge 114, 151
generation-recombination (g-r) noise
 carriers 11, 12, 45
 phonons 17
 superposition of 14, 64, 67
Ge-on-insulator MOSFETs 151

Hanning 35
heterodyne receiver 182
HfAlO$_x$ 145, 156
HfO$_2$ 54, 138, 142, 145, 156
HfSiON 138, 145, 156
high electron mobility transistors (HEMTs) 107
high-frequency noise measurements 40
high-k dielectric materials, table of 138
high-k gate dielectrics *See* CMOS technology
holes 11, 59, 73
homodyne receiver *See* direct-conversion receiver (DCR)
Hooge noise model 15, 17, 78, 82
Hooge parameter 15, 17, 45, 78, 80, 85, 109, 112, 142, 152
hot carriers 64

impact ionization effects 129, 131
independent random variable 20
induced gate noise 40
infrared photon emission 17
inhomogeneous current flow 80
input/output impedance 175
interfacial layer 48, 138, 145, 147
International Technology Roadmap for Semiconductors (ITRS) 106
intrinsic level 56
ionized impurities 59

Index

JFETs 78, 84, 160
Johnson noise *See* thermal noise
junction depth 106

Kirchoff's laws 19

La_2O_3 145
lattice constant 17, 114
lattice defects 83
LC-tank 178
local loading effects 116
Lorentzian 11, 13, 131
low-field mobility 96
low-frequency (LF) noise
    compact noise models 92
    correlated mobility fluctuations 66, 71, 76, 91, 118, 150
    definition 5
    diagnostics 80
    fundamental sources 11, 12, 14
    measurements 28, 36, 39, 41, 48, 144
    mixer 183
    mobility fluctuation noise model 78
    mobility fluctuations subthreshold 78, 82
    MOSFET model 41, 60
    number fluctuation noise model 64
    number fluctuations subthreshold 69
    number fluctuations thermally activated traps 70
    origin of mobility fluctuations 85
    substrate bias effects 86, 91, 154
    switched bias 77
    upconversion 176, 179
low-frequency noise performance
    figure-of-merit 45, 109
    high-k gate dielectrics 90, 141, 158
    metal gate devices 157
    multiple gate MOSFETs 160
    nitrided gate oxide 112
    Schottky Barrier (SB) pMOSFETs 135
    Si MOSFETs 86, 90, 121, 126
    SiGe pMOSFETs 90, 117
    SOI MOSFETs 131
    strained-Si MOSFETs 125
    ultra scaled CMOS devices 110
low-noise amplifier (LNA) 28, 31, 184
low-noise current amplifier 29
low-noise voltage amplifier 29

Matthiessen's rule 59, 80
McWorther model 64, 66, 76
mean free path 17, 87
mean value 3
measurement equipment 31
metal film resistors 31, 39
metal gate *See* under CMOS technology
metal organic chemical vapour deposition (MOCVD) 139
microprocessor 104
mixer 182
Mo 157
mobility fluctuations 7, 15, 16, 43, 64, 77, 78, 84, 85, 95, 119, 147, 153, 155
molecular beam epitaxy (MBE) 139
Moore's law 53
MOSFET
    $1/f$ noise 64, 78
    body-effect coefficient 55, 57, 128
    carrier mobility 59, 73, 88, 96, 115, 123, 133, 138, 141, 153, 159
    channel 18, 42, 54, 60, 82, 84
    channel resistance 31
    depletion region 88, 105, 130
    depletion-layer capacitance 58, 91
    diffusion current 57
    drain current 55
    drain current linear region 55
    drain current noise 60
    drain current saturation region 57
    drain current subthreshold 57
    drain voltage 57
    electron mobility 81, 123
    feature size 104
    flat-band voltage 56
    gate leakage current 137
    gate length 103, 104, 110
    gate oxide 54, 83, 137
    gate oxide thickness 46, 110
    gate voltage 55
    gate voltage overdrive 78
    hole mobility 81, 123
    input gate voltage noise 95
    inversion charge density 55, 65

inverter delay 104
I-V characteristics 58
linear region 63
n-channel 54, 64, 76
noise equivalent circuit 60
noise model 60
off-current 105
oxide capacitance 46, 55
p-channel 54, 64, 71, 76
pinch-off 57
S/D resistance 42, 54, 60
saturation 57
saturation region 63, 78
structure 54
substrate bias 84, 88
subthreshold slope 57
supply voltage 105, 175
thermal noise 62
threshold voltage 55, 56, 88, 105
threshold voltage instability 138, 144
transconductance 46, 62, 89
transition frequency 104, 184
multiple-gate devices *See* under CMOS technology
multiplication factor 64

NiSi 137, 157
nitrided gate oxide 68, 112, 138
noise
    addition of 19
    circuit analysis 18
    circuits 175
    compact models 92
    definition 2
    gate leakage current 113
    in MOSFETs 60
    parameters table of 94
    power 4, 6, 8, 31, 176
    representation of sources 18
    resistance 6
    temperature 6
noise factor *See* noise figure
noise figure 31, 41, 184
noise figure meter 41
nonquasi-static effect 63
normal distribution 3, 4
normalized drain current noise 43
Norton equivalent 18

number fluctuations 7, 14, 16, 43, 64, 77, 80, 84, 95, 118, 132, 155
number of carriers 11, 15, 78, 83
Nyquist noise *See* thermal noise

Omega FETs 130, 159
on-wafer measurements 28
oscilloscope 29, 36
oxide charge 74
oxide trap density 47, 67, 76, 109, 132, 142
oxide/channel interface 46, 59, 68, 72, 84, 87, 113
Oxynitrides *See* nitrided gate oxide

parasitic capacitances 40
parasitic channel 117
partially depleted (PD) SOI 129, 131
passive mixer 183
permittivity 56, 76
phase noise 176, 178, 181
Philips MOS model 11 92
phonon scattering 15, 17, 44, 59, 80, 85, 140
phonon-phonon scattering 18, 87
physical vapour deposition (PVD) 139
Planck's constant 8
pn-junction 9, 18
Poisson process 3, 9, 12, 38
poly-depletion 106, 157
poly-SiGe 157
poly-silicon 54, 108
popcorn noise *See* random-telegraph-signal (RTS) noise
power dissipation 106, 175
power spectral density 3, 4, 5, 28, 29, 36
probability density function 3, 4
probability theory 3
probe 39
PtSi 137
pulse height 12, 38

quality factor 181
quantum correction 8
quantum noise theory 17
quasi-Fermi level *See* Fermi-level

random fluctuations 6
random thermal motion 7

Index

random-telegraph-signal (RTS) noise 3, 12, 38, 45
reactive elements 9
receiver 176
rectangular window 35
reflection 41
relaxation 114, 127
reliability 106
remote phonon scattering 140, 153, 156, 157, 158
resistance fluctuations 15
resistive networks 20
resistor 8
resolution bandwidth 35
RF circuits 53
ring oscillator 181
root mean square 6
Ru 157

scaling rules 103, 106
scattering parameter 46, 65, 71, 73, 76
Schottky Barrier (SB) MOSFETs 135
screening 71, 76
shallow-trench-isolation 125
shielding 29, 39
short-channel effect 103, 105, 159
shot noise 9, 63
Si 114, 151
Si/SiO$_2$ interface 77, 123
SiGe 107, 114, 123, 151
SiGe pMOSFETs *See* under CMOS technology
signal-to-noise ratio 31, 176
silicide 135, 157
silicon permittivity 47
silicon-on-insulator (SOI) *See* under CMOS technology
SIMOX 131
SiO$_2$ 54, 137
skin depth 30
small-signal equivalent circuit 18, 61
source 54
source admittance 41
source impedance 41
source impedance tuner 41
source reflection factor 41
source resistance 32
spectrum analyzer 28, 34
speed 104, 175

SPICE model 92
stationary process 4
strain 114, 123
strong inversion 43, 80, 91
substrate 54
subthreshold 43
superposition principle 20
surface carrier concentration 46
surface potential 47, 56, 65
surface roughness 83, 160
surface roughness scattering 59, 85, 112, 140
symmetry of circuit 181
system bandwidth 9

Ta$_2$O$_5$ 145
TaN 157
TaSiN 157
TCAD simulation 85, 89, 133
temperature dependence 60
temperature sensitivity factor 140
thermal conductivity 87
thermal equilibrium 59
thermal noise 7, 62
thermal velocity 59
Thévenin equivalent 18
threshold voltage *See* under MOSFET
TiAlN 157
time constant 11, 13, 38, 68
time domain 34, 36
time durations 12, 38
TiN 157
transceiver 176
transmission 41
transmission electron microscopy (TEM) 42
trap position 46
trapped charge 59
trapping 13, 68
traps
  acceptor 72
  donor 72
  energy distribution 76
  gate oxide 41, 45, 46, 58, 65, 68, 70, 77, 138, 158, 160
  interface 70, 77, 116
  semiconductors 11, 14
tunneling 66, 70, 77, 106, 129, 135, 142
tunneling attenuation length 68, 144
tunneling distance 46

UNIBOND substrates 128, 131
universal mobility 140
U-shaped distribution 77

valence band 70, 76
variance 3, 11
velocity saturation 63, 184
Wentzel-Kramers-Brillouin (WKB) theory 68
white noise 5, 8, 9, 62

Wiener-Khintchine theorem 4
window function 35
virtual substrates 125
voltage controlled oscillator (VCO) 176, 178
voltage swings 175
volume inversion 131, 159, 160
work function 56

Printed in the United States
84340LV00002B/244-270/A